Walter Krämer · Harald Sonnberger

The Linear Regression Model Under Test

Physica-Verlag Heidelberg Wien

Professor Dr. WALTER KRÄMER, Fachbereich Wirtschaftswissenschaften, Universität Hannover, Wunstorfer Str. 14, D-3000 Hannover 91, FRG

Dr. HARALD SONNBERGER, Leiter der Projektgruppe IAS-SYSTEM, Institut für Höhere Studien, Stumpergasse 56, A-1060 Wien, Austria

ISBN-13: 978-3-642-95878-6   e-ISBN-13: 978-3-642-95876-2
DOI: 10.1007/978-3-642-95876-2

CIP-Kurztitelaufnahme der Deutschen Bibliothek
Krämer. Walter: The linear regression model under test/Walter Krämer; Harald Sonnberger. – Heidelberg; Wien: Physica-Verlag, 1986. ISBN-13: 978-3-642-95878-6

© by Physica-Verlag Heidelberg 1986
Softcover reprint of the hardcover 1st edition   1986

Cover: Erich Kirchner, Heidelberg

7120/7130-543210

PREFACE

This monograph grew out of joint work with various dedicated colleagues and students at the Vienna Institute for Advanced Studies. We would probably never have begun without the impetus of Johann Maurer, who for some time was the spiritus rector behind the Institute's macromodel of the Austrian economy. Manfred Deistler provided sustained stimulation for our research through many discussions in his econometric research seminar. Similar credits are due to Adrian Pagan, Roberto Mariano and Garry Phillips, the econometrics guest professors at the Institute in the 1982 - 1984 period, who through their lectures and advice have contributed greatly to our effort. Hans Schneeweiß offered helpful comments on an earlier version of the manuscript, and Benedikt Poetscher was always willing to lend a helping hand when we had trouble with the mathematics of the tests. Needless to say that any errors are our own.

Much of the programming for the tests and for the Monte Carlo experiments was done by Petr Havlik, Karl Kontrus and Raimund Alt. Without their assistance, our research project would have been impossible. Petr Havlik and Karl Kontrus in addition read and criticized portions of the manuscript, and were of great help in reducing our error rate. Many of the more theoretical results in this monograph would never have come to light without the mathematical expertise of Werner Ploberger, who provided most of the statistical background of the chapter on testing for structural change. We are also indebted to the people at the Institute's computer center (WSR), who, without blinking, watched us use up record amounts of CPU time on their Univac 1100 machine. Beatrix Krones did an expert typing job, and Jörg Breitung and Ottmar von Holtz then edited the manuscript on the Hanover University text processing system. We thank all of them for their help and the effort they have put into our project. Finally we apologize for biting the hands that so generously fed us with data, in particular to John Rea, Dan Nichols and the editorial staff of the Journal of Money, Credit and Banking.

Hanover and Vienna, May 1986   W. Krämer
                                                                         H. Sonnberger

CONTENTS

1. INTRODUCTION

*"The three golden rules of econometrics
are test, test and test."*
D. Hendry

The linear regression model is by far the most widely used means of data description, analysis and prediction in economics and elsewhere. Partly because of this popularity, it is very often used in situations where its assumptions do not apply. Much too often, empirical "laws" are produced, whether unintentionally or by some sort of data mining, which have little to do with reality. As a result, statistical analyses nowadays tend do be either greatly discounted or completely ignored. The golden days of empirical econometrics are definitely over.

As a consequence, recent years have witnessed a remarkable growth of interest in testing - rather than estimating - econometric models. While it took more than a quarter of a century for the first serious article on testing to appear in Econometrica (the Chow test in 1960), the predominance of testing among articles in theoretical econometrics can hardly be overlooked in more recent volumes. In the 1980-1984 period alone, about fifty articles and notes appeared with a focus on testing econometric models.

The present monograph aims to survey and apply some recent contributions in this area. As the title indicates, we mostly confine ourselves to situations where the null hypothesis is that the assumptions of the standard linear regression model apply, though some extensions to dynamic models and simultaneous equation systems will be discussed as well. We will in addition focus on tests for the proper specification of what, in the context of the classical linear regression model, might conveniently be called the "deterministic" part - as opposed to the stochastic part - of the model.

This emphasis is not in line with the relative attention that is given to these problems in the literature. Theoretical econometricians and statisticians have long shown considerable concern for the disturbance structure of regression models. Tests for normality, homoskedasticity and independence of the disturbances as well as procedures to remedy the adverse effects of such departures from the classical assumptions are discussed at length in the literature and in the textbooks. Yet the complications that arise from non well behaved errors seem relatively minor compared to the omission of a relevant independent variable, an incorrect functional form, structural change or correlation between regressors and error terms introduced through simultaneous equations or errors in the variables. While a failure of the stochastic specification in general leads to inefficient, but still consistent parameter estimates, a wrong specification of the deterministic part of the model typically renders parameter estimates inconsistent or even, for lack of a well defined parent parameter, completely meaningless. There is thus a considerable imbalance involved in detecting deviations from the stochastic and deterministic specifications, which is not yet reflected in the attention paid to these problems in applied research.

If one views the latter type of model deviations as affecting the mean, rather than the covariance structure, of the regression disturbances, this can also be phrased in terms of the relative importance of a correct specification of the first as opposed to the second moment of the dependent variable. We think that it matters more to have the mean of the dependent variable correct (conditional on the independent variables), than to be precise about its higher moments.

We discuss the tests for correct specification of the disturbances mainly because of their ability to also detect irregularities in the regressor specification. It is for instance well known that the Durbin-Watson test will also detect incorrect functional form, or that a significant test statistic when checking for normality can also be due to outliers in the data. This is because the null hypothesis for such tests not only maintains that the disturbances are well behaved, but also that they are independent of the regressors, and that the regressors are correctly specified. Failure of any of these conditions can lead to a significant test statistic. Some authors welcome this as an additional merit of their tests, but this lack of robustness is a mixed blessing at best. When the null hypothesis is rejected, one often does not know why. We will briefly return to this important issue in chapter 6.

The plan of the monograph is as follows: Chapter 2 presents some technical preliminaries. We then in chapter 3 survey those procedures we think are less important from an applications viewpoint - tests for independence, homoskedasticity and normality of the disturbances. Tests for the correct specification of the regressors - the deterministic part of the model - are then discussed in chapter 4. Chapter 5 presents some unifying themes that might help to organize this vast material, and chapter 6 applies selected tests to some recent empirical papers from various top level journals. The Appendix finally presents the data, plus the econometric software package that we were using for our analysis (the IAS-SYSTEM).

2. TECHNICAL PRELIMINARIES

a) The Linear Regression Model

This section mainly introduces some common notation, and is not intended as a text-book introduction to linear regression. Convenient references for this purpose are *Koerts* and *Abrahamse* (1969), *Theil* (1971), or many other intermediate level statistics and econometrics textbooks.

Notation and assumptions

The point of departure for most of what follows, in this and subsequent chapters, is the standard (classical) linear regression model

$$y_t = x_t'\beta + u_t \quad (t = 1,...,T), \tag{2.1}$$

where y_t denotes the t'th observation of the endogenous (dependent-, response-) variable, and x_t (Kx1) is a column vector of observations on K exogeneous (independent-, regressor-, design-) variables. The index t stresses the fact that we will mostly be concerned with time series data. β is the Kx1 vector of unknown regression coefficients to be estimated, and u_t is a stochastic, unobservable disturbance term (sometimes also called latent variable or error in the equation).

The familiar matrix notation for this model is

$$y = X\beta + u. \tag{2.2}$$

The assumptions which constitute the linear regression model are

A1: X is nonstochastic, with column rank K,

A2: u is a random vector with expectation zero,

A3: $\text{cov}(u) = E(uu') = \sigma^2 I$ (i.e. the u_t's are uncorrelated and have identical variance σ^2), and

A4: the u_t's are jointly normal.

Assumption A1 explicitly excludes lagged dependent variables among the regressors or simultaneous equation models. Since these have important applications in econometrics and elsewhere, we will however allow for stochastic regressors at some instances below. Assumption A2 mainly requires that none of the nonincluded variables captured by u should exert a systematic influence on y. Assumption A3 maintains that the u_t's are both serially uncorrelated and homoskedastic, and assumption A4 (which is not needed for many results) further specifies the distribution of the disturbances. In what follows, these assumptions will always constitute the null hypothesis of the various tests, unless explicitly stated otherwise.

Very important additional assumptions implicit in (2.2) are that the regression parameters do not change over time, that there are no measurement errors in X, that all relevant regressors are indeed included in the equation, and that the relationship between the dependent and the independent variables is indeed of a linear form. None of these assumptions is obvious in most applications, and the tests to be discussed in this monograph are supposed to determine whether in a given situation they hold or not.

Whenever it comes to large sample results, one additional assumption is needed on the limiting behaviour of $X'X$. We follow standard practice by assuming

$$\lim_{T \to \infty} T^{-1} X'X = Q \tag{2.3}$$

for some nonsingular KxK matrix Q.

It might however be worth noting that (2.3) excludes trended data. For $X_t = t$, one for instance has $X'X = O(T^3)$, i.e. $T^{-1}X'X \to \infty$, and many standard proofs do not carry over to such cases. *Krämer* (1984a, 1984b, 1985a) provides a discussion of the effect of trended data on large sample properties of econometric estimators.

The Ordinary Least Squares (OLS) estimate for β in the model (2.2) is

$$\hat{\beta} = (X'X)^{-1} X'y. \tag{2.4}$$

Direct computation shows that $E(\hat{\beta}) = \beta$ and

$$\text{cov}(\hat{\beta}) = \sigma^2 (X'X)^{-1}. \tag{2.5}$$

It is well known that $\hat{\beta}$ is BLUE (Best Linear Unbiased Estimator), given assumptions A1-A3, i.e. no other linear unbiased estimator has a covariance matrix smaller than (2.5). If assumption A4 also holds, $\hat{\beta}$ is optimal also in the larger class of merely unbiased estimators.

In addition, (2.3) implies that $X'X^{-1} \to 0$ as $T \to \infty$, that is $\text{cov}(\hat{\beta}) \to 0$, and $\hat{\beta}$ is (weakly) consistent.

Failure of any of the assumptions A1-A4 and the assumptions implicit in (2.2) has different effects for the performance of $\hat{\beta}$. If $\text{cov}(u) = V$, where $V \neq \sigma^2 I$, $\hat{\beta}$ is still unbiased, and under mild conditions on V also consistent, but its covariance matrix is now

$$\text{cov}(\hat{\beta}) = (X'X)^{-1}X'VX(X'X)^{-1}. \tag{2.6}$$

In general, (2.6) will differ from (2.5) (*Neuwirth* (1982) gives conditions for these matrices to be identical, but his requirements will almost never be met in practice). One will therefore produce wrong inferences when proceeding as if $V = \sigma^2 I$. Also, there are more efficient linear estimators when $V \neq \sigma^2 I$. Unpleasant as these consequences may be, however, it seems fair to say that empirical econometrics would be in much better shape today, and would probably not experience any crisis at all, if these were the only problems pleaguing applied work.

Producing a wrong probability statement due to an incorrect disturbance covariance matrix indeed seems a minor affair as compared to the effect of e.g. omitting a relevant regressor, structural change or specifying the wrong functional form. Usually, $\hat{\beta}$ will cease to be consistent, or might even become completely meaningless for lack of a well defined parent parameter. A well known case in point is the inconsistency of OLS in the context of correlation between regressors and disturbances, as introduced by e.g. errors in variables or simultaneous equations. Or else, if β changes during the sample period, it is not clear at all what is estimated by $\hat{\beta}$. We will therefore in this monograph stress those tests that help detect failures in the assumptions on the nonstochastic part of the model. Though such failures may also be classified as affecting the disturbances - by producing a nonzero mean -, we prefer to speak in terms of the stochastic and nonstochastic parts of the model, respectively.

Regression residuals

Many tests discussed below are based on the residuals from an OLS regression. Denoting this vector by $\hat{u}$, we have

$$\hat{u} = y - X\hat{\beta} = y - X(X'X)^{-1}X'y = My , \qquad (2.7)$$

where

$$M = I - X(X'X)^{-1}X'. \qquad (2.8)$$

This matrix will reappear at various stages below. It is easily checked that M is idempotent and of rank T-K, and that $MX = 0$ (i.e. M is the orthogonal projection on the orthogonal complement of the column space of X). Thus,

$$\hat{u} = My = M(X\beta + u) = Mu , \qquad (2.9)$$

i.e. the OLS-residuals are linear combinations of the unobservable disturbances. We will use $\hat{u}$ to fill in for u at various places below.

From (2.9),

$$\hat{u} - u = (M-I)u = -X(X'X)^{-1}X'u, \qquad (2.10)$$

which implies that $\text{cov}(\hat{u}-u) = \sigma^2 X(X'X)^{-1}X'$. (2.3) in addition implies that the maximum element in $X(X'X)^{-1}X'$ tends to zero as $T \to \infty$, so

$$\hat{u}_t - u_t \xrightarrow{p} 0 \quad \text{uniformly for all t.} \qquad (2.11)$$

We can therefore use $\hat{u}$ as an increasingly good approximation for u as T increases. (2.11) further implies that any element of $\hat{u}$ tends in distribution to the corresponding element of u.

However, from (2.9),

$$\text{cov}(\hat{u}) = \sigma^2 M, \tag{2.12}$$

i.e. the elements of $\hat{u}$ are neither uncorrelated nor homoskedastic, even when the model is correctly specified. This puts a considerable obstacle in the way of constructing test statistics based on $\hat{u}$.

There are several ways around this dilemma. In general, any residual vector which is linear in the y_t's may be written as

$$u^* = Cy,$$

where the matrix C should satisfy

$$CX = 0 \tag{2.13}$$

in order that the residual vector has the same zero expectation as the corresponding vector of true disturbances (this follows from the fact that $E(Cy) = E[C(X\beta + u)] = CX\beta = 0$ for all β if and only if $CX = 0$). Then $u^* = Cy = Cu$, with covariance matrix

$$E(Cuu'C') = \sigma^2 CC'.$$

Requiring $CC' = I$ irrespective of X produces the important class of LUS (Linear Unbiased Scalar covariance matrix) residuals. These residuals are $NID(0,\sigma^2)$ under correct specification. The restrictions on C here imply that C can have at most T-K rows, i.e. that at least K disturbances will have to be disregarded when one insists on an unbiased residual vector with a scalar covariance matrix. When X has full column rank, as we are assuming throughout, one can in addition always find a C matrix with exactly T-K rows. $u^* = Cu$ is then interpreted as an approximation for a T-K subvector $u^{(1)}$ of u.

In general there are infinitely many such C matrices for any given X. *Theil* (1965) has suggested to choose C such that

$$E[(u^{(1)}-u^*)'(u^{(1)}-u^*)] \tag{2.14}$$

is minimized, i.e. such that the u_t^*'s approximate the corresponding u_t's as closely as possible (in the mean square error sense). The resulting residuals are then called BLUS (Best LUS) residuals.

BLUS-residuals are still not unique, since they depend on the subvector $u^{(1)}$ of u that was selected for approximation. There are $\binom{T}{K}$ such choices, and the most appropriate one will in general depend on the purpose for which the residuals are to be used. When testing against autocorrelation, successive disturbances seem most reasonable, while discarding disturbances in the middle of the sample seems advisable when testing against heteroskedasticity.

Since we will not be using BLUS-residuals later, there is no point in discussing the quite substantial computational difficulties involved. Convenient references are *Koerts* and *Abrahamse* (1969, chapter 3.5) or *Theil* (1971, chapter 5.2).

We will however make heavy use in what follows of another class of LUS-residuals called recursive residuals. These have been introduced independently by several authors. Our exposition follows *Phillips* and *Harvey* (1974) and *Brown* et al. (1975).

Let

$$\hat{\beta}^{(t)} = (X^{(t)} {}'X^{(t)})^{-1} X^{(t)} {}'y^{(t)} \tag{2.15}$$

be the OLS-estimate for β based on the first t observations. In (2.15), $X^{(t)}$ comprises the first t rows of X, and $y^{(t)}$ (not to be confused with y_t, the t'th observation on y) contains the first t elements of y. The invertibility of $X^{(t)} {}'X^{(t)}$ requires that $t \geq K$, and we will in particular assume that $X^{(t)} {}'X^{(t)}$ is nonsingular for $t = K$ and thus also for all $t > K$ (This is not so innocent an assumption as might seem on first sight, in particular when dummy variables are present among the regressors, and we will go at some length later to discuss what can be done if $X^{(t)} {}'X^{(t)}$ has rank deficiences for some $t > K$).

The recursive residuals $\tilde{u}_t$ are now for $t = K+1,...,T$ defined by

$$\tilde{u}_t = \frac{y_t - x_t{}'\hat{\beta}^{(t-1)}}{(1 + x_t'(X^{(t-1)}{}'X^{(t-1)})^{-1}x_t)^{1/2}} \tag{2.16}$$

This formula also demonstrates that the recursive residuals are just standardized forecast errors, where the normalization factor

$$f_t = (1 + x_t'(X^{(t-1)}{}'X^{(t-1)})^{-1}x_t)^{1/2} \tag{2.17}$$

ensures that all $\tilde{u}_t$'s $(t = K+1,...,T)$ have the same variance σ^2. The $(T-K) \times T$ C-matrix implicit in the computation of $\tilde{u}$ has t'th row

$$[-x_t{}'(X^{(t-1)}{}'X^{(t-1)})^{-1}x_t f_t^{-1}, f_t^{-1}, 0, ..., 0]. \tag{2.18}$$

It can therefore be easily shown by direct multiplication that $CX = 0$ and $CC' = I$, i.e. that (2.16) indeed produces LUS-residuals. By definition, Theil's BLUS-residuals u^* approximate u better than $\tilde{u}$, but the $\tilde{u}_t$'s depend only on u_i's with $i \leq t$ and are therefore more useful for certain types of specification tests.

They do however *not* tend in distribution to the true disturbances as $T \to \infty$, in contrast to the OLS and BLUS residuals. It is immediate from (2.16) that $\tilde{u}_t$ does (for $T > t$) not depend on T and is therefore not affected when sample size increases. In particular, $\tilde{u}_t$ cannot approximate u_t any better as $T \to \infty$.

Rather than requiring that the covariance matrix of u^* be scalar, any other type of fixed covariance structure for the residuals could do as well. The point with LUS residuals is not that the residuals are independent, but rather that their correlation structure does not depend on the X-matrix. Such procedures are surveyed by *Dubbelman* (1978) or *King* (1983a, section 5) and will be neglected here.

b) LR-, Wald- and LM-Tests

Many of the tests below are derived, whether explicitly or implicitly, from either the Likelihood Ratio (LR)-, Wald-, or Lagrange Multiplier (LM)-principle. These procedures are well known and widely discussed in the statistics literature and provide a convenient common framework for otherwise quite dissimilar tests (other unifying themes will be discussed in chapter 5). Again, in this section, we mainly establish some common notation, without getting lost in technical intricacies. Hopefully, this will also help to better appreciate the rationale for some of the tests presented later. The interested reader is referred to *Breusch* and *Pagan* (1980), *Harvey* (1981, chapter 5) or *Engle* (1984) for details. A very useful pedagogical introduction to the subject is *Buse* (1982).

Basic principles

Let in general y be a $T \times 1$ random vector drawn from a joint density $f(y,\theta)$, where θ is an unknown $K \times 1$ vector of parameters, from some set $\Theta \subseteq \mathbb{R}^K$.

Throughout we assume that f and θ satisfy standard regularity conditions (which basically call for the possibility of a two term Taylor series expansion, and the interchangeability of differentiation and integration). For our purposes it will be sufficient to remember that these regularity conditions always hold in the applications below.

Let the hypothesis under test be that $\theta \in \Theta_0 \subseteq \Theta$. Usually, Θ_0 is defined by $R(\theta) = 0$, where R is some $p \times 1$ vector of functions defined on Θ. More often than not, R is in addition such that θ may be partitioned into $\theta = [\theta_1', \theta_2']'$, and the null hypothesis is that $\theta_1 = 0$, whereas θ_2 is unrestricted.

The Maximum Likelihood (ML) estimate $\hat{\theta}$ for θ maximizes $f(y,\theta)$, viewed as a function of θ, given y. $\hat{\theta}$ then also maximizes the log likelihood

$$L(\theta,y) = \log f(y,\theta) , \tag{2.19}$$

and vice versa. The main reason for arguing in terms of L rather than f is that taking logs transforms products into much better manageable sums.

Under the usual regularity conditions, $\hat{\theta}$ satisfies

$$\frac{\partial L}{\partial \theta}(\hat{\theta},y) = 0 , \tag{2.20}$$

i.e. $\hat{\theta}$ sets the score

$$S(\theta,y) = \frac{\partial L}{\partial \theta}(\theta,y) \tag{2.21}$$

to zero. Let in addition $\bar{\theta}$ be the ML-estimate for θ obtained under the restriction that $\theta \in \Theta_0$, and let $S(\bar{\theta},y)$ be the score evaluated at $\bar{\theta}$. $S(\bar{\theta},y)$ will in general be different from zero. Finally, define a $K \times K$ matrix $I(\theta)$ by

$$I(\theta) = -E \frac{\partial^2 L}{\partial \theta \partial \theta'}(\theta) . \tag{2.22}$$

$I(\theta)$ is known as Fisher's information matrix . It is positiv definite under the usual regularity conditions.

The Likelihood Ratio test is now based upon the difference between the maximum of the log likelihood under the null and under the alternative hypotheses: If this difference is too large in absolute value, the null hypothesis is rejected (the term "ratio" appears because this is equivalent to rejecting the null hypothesis whenever the ratio of the respective likelihoods is too small). The appropriate critical value is determined from the well known fact that under H_0, the LR-statistic

$$\xi_{LR} = -2(L(\bar{\theta},y)-L(\hat{\theta},y)) \tag{2.23}$$

has a limiting χ^2 distribution with p degrees of freedom.

The Wald test relies on $R(\hat{\theta})$. This should be close to zero if H_0 is true. In particular, if R is linear (i.e. a $p \times K$ matrix), it appears reasonable to reject H_0 whenever

$$\xi_w = \hat{\theta}'R'[RI(\theta)^{-1}R']^{-1}R\hat{\theta} \tag{2.24}$$

is too large. The appropriate critical value is again obtained by observing that ξ_w has a limiting χ_p^2 distribution under H_0.

The Lagrange-Multiplier principle finally rests on the idea that if the restrictions are valid, $\bar{\theta}$ will be close to $\hat{\theta}$, and so $S(\bar{\theta},y)$, the vector of partial derivatives of the log likelihood function, will be close to zero. We therefore reject H_0 whenever

$$\xi_{LM} = S'(\bar{\theta},y)I(\theta)^{-1}S(\bar{\theta},y) \tag{2.25}$$

is too large, where the test statistic can again be shown to be asymptotically χ_p^2 (the term "Lagrange Multiplier test" derives from the fact that $S(\theta)$ can be expressed in terms of the Lagrange Multipliers obtained from the restricted maximization of the log likelihood).

The three principles are thus based on different statistics which measure the plausibility of H_0. The LR test is formulated in terms of $L(\bar{\theta})-L(\hat{\theta})$, the Wald test in terms of $R(\hat{\theta})$ and the LM test in terms of $S(\bar{\theta})$.

Consider for illustration the case $K=1$ and H_0: $\theta = \theta_0$, i.e. a single unknown parameter and a simple null hypothesis. Here we obviously have $\bar{\theta} = \theta_0$. This situation is depicted in Figure (2.1), where we plot the log likelihood function against θ for some particular realization of y. The LR test is based on the vertical difference between $(\bar{\theta},L(\bar{\theta}))$ and $(\hat{\theta},L(\hat{\theta}))$, the Wald test is based upon the horizontal difference, and the LM test is based on the slope of the log likelihood function at $(\bar{\theta},L(\bar{\theta}))$, where each appears to be a reasonable measure of the extend to which the data agree with H_0.

FIG. 2.1: LOG-LIKELIHOOD FUNCTION WITH SIMPLE NULL HYPOTHESIS

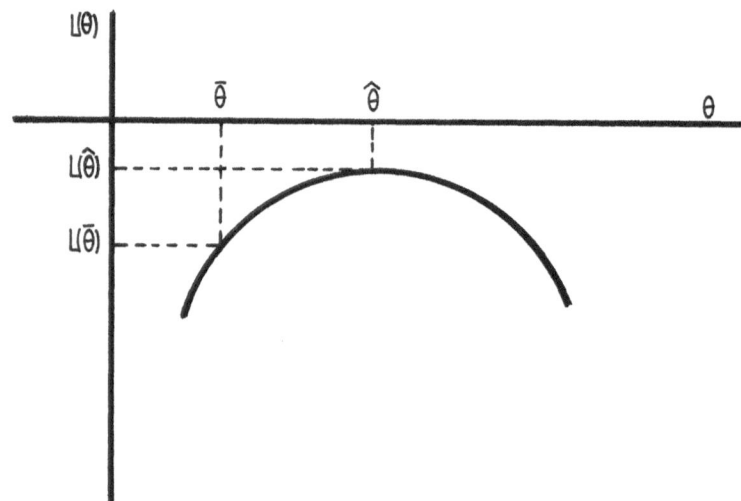

A simple example

As an additional example, which may also serve as an introduction to more complicated cases in subsequent chapters, consider the problem of testing for the validity of a set of linear restrictions for the regression coefficients in the standard linear regression model (2.2). There is the well known F-test available here, and so let us see how this procedures relates to the LR-, Wald- and LM tests of the same hypothesis. For simplicity, we assume that the disturbance variance σ^2 is known, i.e. that the vector θ of unknown parameters is identical to the regression coefficient vector β. Let the null hypothesis be that

$$R\beta = 0 ,$$

where R is a known and fixed $p \times K$ matrix $(p < K)$ with linear independent rows.

Under the assumptions of section 2.a, the joint density for the y_t's in the model (2.2) is

$$f(y,\cdot\beta) = 2\pi^{-T/2}\sigma^{-T}\exp\{-\frac{1}{2\sigma^2} \ (y-X\beta)\,'(y-X\beta)\} . \tag{2.26}$$

The log likelihood therefore is

$$L(\beta,y) = (-T/2)\log 2\pi - T \log \sigma - \frac{1}{2\sigma^2}(y-X\beta)\,'(y-X\beta), \tag{2.27}$$

where we will in what follows abbreviate $(-T/2) \log 2\pi - T \log \sigma$ by K. The score function is therefore given by

$$S(\beta,y) = \frac{\partial L}{\partial \beta} = \frac{1}{\sigma^2} X'y - \frac{1}{\sigma^2} X'X\beta \;, \tag{2.28}$$

which in turn implies that

$$\frac{\partial^2 L}{\partial \beta \partial \beta'}(\beta) = -\frac{1}{\sigma^2} X'X \tag{2.29}$$

Since (2.29) does not depend on the observations in y, we in addition immediately deduce that

$$I(\theta) = -E \frac{\partial^2 L}{\partial \beta \partial \beta'}(\beta) = \frac{1}{\sigma^2} X'X \;. \tag{2.30}$$

In order to construct the LR statistic we need both the restricted and the unrestricted ML estimates for β, plus the log likelihood at these values. Setting (2.28) to zero produces the unrestricted ML estimate $\hat{\beta} = (X'X)^{-1}X'y$. This is simply the OLS-estimate for β. The restricted ML estimate is also well known to be (see e.g. *Theil*, 1971, p. 285)

$$\bar{\beta} = \hat{\beta} - (X'X)^{-1} R'(R(X'X)^{-1}R')^{-1}R\hat{\beta} \tag{2.31}$$

From (2.31) it is clear that $\hat{\beta}$ and $\bar{\beta}$ will be identical whenever the unrestricted estimate obeys the restrictions imposed by the null hypothesis, i.e. when $R\hat{\beta} = 0$.

Defining $\hat{u} = y\text{-}X\hat{\beta}$ as usual and $\bar{u} = y\text{-}X\bar{\beta}$ we can now write the unrestricted log likelihood as

$$L(\hat{\beta},y) = K - \frac{1}{2\sigma^2} \hat{u}'\hat{u} \tag{2.32}$$

and the restricted log likelihood as

$$L(\bar{\beta},y) = K - \frac{1}{2\sigma^2} \bar{u}'\bar{u} \tag{2.33}$$

The LR statistic is therefore

$$\xi_{LR} = -2(L(\bar{\beta},y)\text{-}L(\hat{\beta},y)) = (\bar{u}'\bar{u}\text{-}\hat{u}'\hat{u})/\sigma^2 \;, \tag{2.34}$$

which has an exact χ_p^2 distribution under H_0. The expression (2.34) resembles the familiar F-statistic for the general case where σ^2 is estimated. Replacing σ^2 by $s^2 = \hat{u}'\hat{u}/(T\text{-}K)$ immediately shows that the F-statistic

$$\xi_F = \frac{(\bar{u}'\bar{u}\text{-}\hat{u}'\hat{u})/p}{\hat{u}'\hat{u}/(T\text{-}K)} \tag{2.35}$$

differs from ξ_{LR} only by the scale factor $1/p$.

Since

$$\bar{u} = y - X\bar{\beta} = \hat{u} - X(X'X)^{-1}R'(R(X'X)^{-1}R')^{-1}R\hat{\beta} \tag{2.36}$$

and

$$\bar{u}'\bar{u} = \hat{u}'\hat{u} + \hat{\beta}'R'(R(X'X)^{-1}R')^{-1}R\hat{\beta} , \tag{2.37}$$

(2.34) can also be expressed as

$$\xi_{LR} = \hat{\beta}'R'(R(\frac{1}{\sigma^2}X'X)^{-1}R')^{-1}R\hat{\beta} . \tag{2.38}$$

This however is identical to the Wald statistic (2.24), as applied to the present problem, since here

$$I(\hat{\beta}) = \frac{1}{\sigma^2}X'X.$$

To obtain the LM statistic, we note that from (2.28), and after substituting (2.31) for β,

$$S(\bar{\beta},y) = \frac{1}{\sigma^2}R'(R(X'X)^{-1}R')^{-1}R\hat{\beta} . \tag{2.39}$$

Thus the general formula (2.25) here again reduces to

$$\xi_{LM} = \hat{\beta}'R'(R(\frac{1}{\sigma^2}X'X)^{-1}R')^{-1}R\hat{\beta}, \tag{2.40}$$

which is identical to ξ_{LR} and ξ_W.

In practice, of course, σ^2 must also be estimated in addition to β, and ξ_{LR}, ξ_W and ξ_{LM} will in general differ due to different estimates for σ^2. It can however be shown, both for the present and more general models, that under H_0 the respective differences tend to zero as $T \to \infty$.

The above example highlights some important properties of the LR-, Wald and LM approaches to hypothesis testing. First, they may lead to well known tests that have been derived along entirely different lines. We will rediscover this phenomenon at various instances below. Second, they tend to produce similar test statistics in spite of the dissimilarity of the three approaches. Since the tests also have similar power characteristics, any choice among them will therefore usually be based on computational convenience.

LR tests in general involve the biggest computational burden, since the model must be estimated both under the null and alternative hypothesis. Wald tests require only estimation of the unrestricted model, whereas the estimation of the restricted model is needed for LM tests. This also explains the tremendous recent popularity of LM tests in econometrics, since in many applications the estimation of the restricted model is almost trivial. This will become apparent when we discuss some particular tests in more detail below.

Estimating the information matrix

A major problem in many applications of the LM principle is the evaluation of the information matrix $I(\theta)$ (see 2.22). $I(\theta)$ was known in the simple regression example above but must be estimated from the data in most real world applications. Depending upon the particular estimator $\hat{I}$ for $I(\theta)$, it is therefore possible to generate a whole class of different test statistics, which are asymptotically equivalent (have the same limiting distribution under the null hypothesis and local alternatives) provided

$$\text{plim}_{T \to \infty} \frac{1}{T} \hat{I} = \lim_{T \to \infty} \frac{1}{T} I(\theta) = \tilde{I} \text{ (nonsingular).} \tag{2.41}$$

A particularly convenient way of obtaining estimates for $I(\theta)$ is based on the fundamental equality

$$I(\theta) = -E\left(\frac{\partial^2 L}{\partial \theta \partial \theta'} \right) = E\left[\left(\frac{\partial L}{\partial \theta} \right)\left(\frac{\partial L}{\partial \theta} \right)' \right], \tag{2.42}$$

where again all expectations and derivatives are evaluated at the true parameter values.

In order to exploit the relationship (2.42), we first rewrite the log likelihood function $L(\theta)$ as

$$L(\theta) = \Sigma_{t=1}^{T} l_t(\theta), \tag{2.43}$$

where $l_t(\theta)$ is the log likelihood for the t'th observation. Now, under H_0,

$$\tilde{I} = \lim_{T \to \infty} \frac{1}{T} I(\theta)$$

$$= -\text{plim}_{T \to \infty} \frac{1}{T}\left(\frac{\partial^2 L}{\partial \theta \partial \theta'} \right) \tag{2.44}$$

$$= \text{plim}_{T \to \infty} \frac{1}{T} \Sigma_{t=1}^{T}\left(\frac{\partial l_t}{\partial \theta}(\bar{\theta}) \right)\left(\frac{\partial l_t}{\partial \theta}(\bar{\theta}) \right)'.$$

(see e.g. *Berndt* et al. 1974), where $\bar{\theta}$ is as before the restricted ML estimate for θ. We may therefore in the formula (2.25) of the general LM test statistic replace $I(\theta)$ by

$$\Sigma_{t=1}^{T}\left(\frac{\partial l_t}{\partial \theta}(\bar{\theta})\right)\left(\frac{\partial l_t}{\partial \theta}(\bar{\theta}) \right)'. \tag{2.45}$$

If $W(\theta)$ is the T by K matrix with typical element $w_{tk} = \partial l_t(\theta)/\partial \theta_k (t=1,...,T; k=1,...,K)$, and i is a T dimensional vector with each element equal to unity, then (2.25) can be rewritten as

$$\xi_{LM} = i'W(\bar{\theta})[W(\bar{\theta})'W(\bar{\theta})]^{-1}W(\bar{\theta})'i \tag{2.46}$$

$$= T[i'W(\bar{\theta})[W(\bar{\theta})'W(\bar{\theta})]^{-1}W(\bar{\theta})'i/i'i]$$

$$= T R^2_{i,\overline{w}},$$

where $R^2_{i,\overline{w}}$ is the coefficient of determination of the regression of i on $W(\bar{\theta})$.

The form (2.46) of the LM-test is by far the most widely used in econometric applications, as will be seen in various examples below.

c) Miscellaneous Terminology

We have already introduced some jargon in section b. Anybody who is utterly unfamiliar with this should stop reading this book. Throughout we assume some basic familiarity with the general statistical test methodology, on a level required for graduate courses in econometrics in most economics departments. Similar to section a, the following remarks are mainly meant to clarify the notation.

Let Θ be the set of unknown parameters for some statistical model (in the linear regression model (2.1), with unknown parameters β and σ^2, $\Theta = \mathbb{R}^K \times \mathbb{R}^+$). Any nonempty subset $\Theta_0 \subset \Theta$ can then define a *null hypothesis* (H_0). The resulting complement $\Theta_1 = \Theta \setminus \Theta_0$ is called the *alternative hypothesis* (H_1).

Let y be a $T \times 1$ random vector whose probability distribution is indexed by $\theta \in \Theta$. A *statistical test* of H_0 against H_1 is then defined by some measurable subset $R \subset \mathbb{R}^T$. If $y \in R$, we reject H_0, and we do not reject H_0 (accept H_0, reject H_1) if $y \in A = \mathbb{R}^T \setminus R$. R is called the *rejection region* and A is called the *acceptance region* of the test. Rejecting H_0 when in fact $\theta \in \Theta_0$ is called a *Type I error* and a *Type II error* is committed whenever we do not reject H_0 when in fact $\theta \in \Theta_1$.

The probability measure of the rejection region R of course depends on $\theta \in \Theta$. The quantity

$$\alpha = \sup_{\theta \in \Theta_0} P_\theta(R) \tag{2.47}$$

is called the *significance level* or the *size* of the test. The test is called *similar* if $P_\theta(R) = \alpha$ for all $\theta \in \Theta_0$, and it is called *unbiased* if $P_\theta(R) \geq \alpha$ for all $\theta \in \Theta_1$.

The size α of the test is typically fixed a priori. The problem is then to find a rejection region R such that (2.47) holds. A test is then called *exact* or *valid* if (2.47) holds exactly. Often however the problem is solved only imperfectly (in fact, this will be almost the rule for the procedures that are discussed in this monograph). A test is then called *conservative* if

$$\sup_{\theta \in \Theta_0} P_\theta(R) < \alpha,$$

and it is called *asymptotic* if

$$\lim_{T \to \infty} \sup_{\theta \in \Theta_0} P_\theta(R) = \alpha . \tag{2.48}$$

Most tests in econometrics are only asymptotic tests.

Strictly speaking, the rejection region R in (2.48) should have been indexed by T (being a subset of $\mathbb{R}^T$), but we often omit this subscript in order not to clutter the notation.

Often the rejection region R is implicitly defined by

$$S(y) \geq c ,\tag{2.49}$$

where S(y) is some scalar function of the observations y and c is some real constant. The interval defined by (2.49) is then also called the rejection region of the test, and we do not distinguish in notation between this interval and R (which is a subset of $\mathbb{R}^T$).

Let $s = S(y)$ be the value of S(y) that corresponds to some given sample y. The quantity

$$\alpha = \text{sub}_{\theta \in \Theta_0} P_\theta(S(y) \geq s)\tag{2.50}$$

is called the *prob-value* of the sample s. It equals the smallest significance level that would have led to a rejection of H_0.

Obviously, we would like $P_\theta(S(y) > c)$ to be large whenever $\theta \in \Theta_1$. A test is called *consistent* if for any given $\theta \in \Theta_1$

$$\lim_{T \to \infty} P_\theta(S \geq c) = 1 .\tag{2.51}$$

Consistency however is only a weak requirement which is satisfied by almost all tests below.

There are various ways to refine this rather indiscriminating concept. The one that has gained the widest acceptance in econometrics is the concept of *local power*. Computation of local power requires that alternatives can be indexed such as to approach the null hypothesis (in some sense) as $T \to \infty$. As an example, consider the F-test of H_0: $\beta = \beta_0$ in the linear regression model. Here a *local alternative* is a sequence

$$\beta_T = \beta_0 + \frac{1}{\sqrt{T}} \Delta\beta,\tag{2.52}$$

of regression coefficients, where $\Delta\beta$ is fixed. Ceteris paribus, an increase in sample size T tends to increase the power of the test. This increase is however balanced by a move towards the null hypothesis of the alternative, so that the power of the test will in general approach a well defined limit in between α and 1 as $T \to \infty$. This limit is then called the *local power* of the test.

A very competent discussion of this and related concepts of asymptotic efficiency of statistical tests is in *Serfling* (1980, chapter 10).

Local power comes into play whenever there are several competing tests. Let R_1 and R_2 be the rejection regions of test 1 and test 2, respectively, both of size α. The tests are called *equivalent* if $R_1 = R_2$ (which is often not obvious at all for tests which are defined via their respective test statistics). The tests are called *asymptotically equivalent* if

$$\lim_{T \to \infty} P_{\theta_T}(R_1) = \lim_{T \to \infty} P_{\theta_T}(R_2) \tag{2.53}$$

for any sequence $\{\theta_T\}$ of local alternatives.

Test 1 is called *uniformly more powerful* than test 2 if both have identical size and if for given sample size T,

$$P_\theta(R_1) \geq P_\theta(R_2) \tag{2.54}$$

for all $\theta \in \Theta_1$. Test 1 is called *uniformly most powerful* if (2.54) holds with respect to all other tests. Usually, there does not exist a uniformly most powerful test.

3. TESTING DISTURBANCES

In this chapter we examine whether the disturbances in the regression model (2.1) are well behaved. As we have noted before, this can also be viewed as a test of whether the higher moments of the dependent variable conform to the assumptions of the model.

The possible deviations from the ideal model assumptions are tackled one at a time. Only little is known on how to proceed when several problems occur simultaneously. This issue will be briefly discussed in chapter 6.

a) Autocorrelation

Testing for autocorrelation might well be the most intensely researched statistical problem in all of econometrics. Convenient surveys are *King* (1983a) or *Judge et. al.* (1980, chapter 5.4). Below we focus on the more popular procedures and on tests that also have power against misspecification alternatives.

The Durbin-Watson test

Despite heavy competition, the most common procedure is still the Durbin-Watson test. It is based on the assumption that the u_t's in (2.1) follow a stationary first-order autoregressive process, i.e. that

$$u_t = \varrho u_{t-1} + \epsilon_t , \quad (t=2,...,T) \tag{3.1}$$

where the ϵ_t's are $\mathrm{NID}(0,\sigma_\epsilon^2)$. Stationarity further implies that all disturbances have common variance σ_u^2, with

$$\sigma_u^2 = (1-\varrho^2)^{-1}\sigma_\epsilon^2, \tag{3.2}$$

and that $|\varrho| < 1$.

The covariance matrix of u may be written as $\sigma_u^2 V$, where

$$V = \begin{bmatrix} 1 & \varrho & & \varrho^{T-1} \\ \varrho & 1 & \ddots & \varrho^{T-2} \\ \vdots & & \ddots & \vdots \\ \varrho^{T-1} & \varrho^{T-2} & & 1 \end{bmatrix} . \tag{3.3}$$

The null hypothesis of no serial correlation is therefore equivalent to $H_0 : \varrho = 0$. The DW test statistic is

$$\begin{aligned} d_1 &= \Sigma_{t=2}^{T} (\hat{u}_t - \hat{u}_{t-1})^2 / \Sigma_{t=1}^{T} \hat{u}_t^2 \\ &= \hat{u}'A\hat{u} / \hat{u}'\hat{u} , \end{aligned} \tag{3.4}$$

where

$$A = \begin{bmatrix} 1 & -1 & & 0 \\ -1 & 2 & \ddots & \\ & \ddots & \ddots & -1 \\ 0 & & -1 & 1 \end{bmatrix} . \tag{3.5}$$

It is easily checked that $0 \le d_1 \le 4$. With positive serial correlation among the u_t's, neighbouring u_t's and thus $\hat{u}_t$'s will tend to be close to each other, i.e. d_1 will tend to be small. On the other hand, when ϱ is negative, d_1 will be large. A plausible two-sided test therefore rejects H_0 whenever d_1 moves too far away from 2, the mid-point of its range. One-sided tests will reject H_0 whenever d_1 is too small (H_1: $\varrho > 0$) or too large (H_1: $\varrho < 0$).

The theoretical foundation for this rule dates back to *Anderson* (1948). He showed that whenever the disturbance vector u has densitiy

$$f(u) = K \exp[-\frac{1}{2\sigma^2}((1+\varrho^2)u'u - 2\varrho u'\phi u)] , \tag{3.6}$$

where K is some constant and ϕ is a symmetric $T \times T$ matrix, a uniformly most powerful (UMP) test of H_0: $\varrho = 0$ against H_1: $\varrho > 0$ is given by

$$\hat{u}'\phi\hat{u}/\hat{u}'\hat{u} > k , \tag{3.7}$$

provided the columns of X are linear combinations of eigenvectors of ϕ.

This result is linked to the DW test as follows: The inverse of cov(u) is

$$\frac{1}{\sigma_u^2}V^{-1} = \frac{1}{\sigma_\epsilon^2}\begin{bmatrix} 1 & -\varrho & & 0 \\ -\varrho & 1+\varrho^2 & \ddots & \\ & \ddots & \ddots & \ddots \\ & & 1+\varrho^2 & -\varrho \\ 0 & & -\varrho & 1 \end{bmatrix} \tag{3.8}$$

$$= \frac{1}{\sigma_\epsilon^2}[(1+\varrho^2)I - 2\varrho\phi - \varrho(\varrho-1)C] ,$$

where

$$
\phi = \frac{1}{2} \begin{bmatrix} 1 & 1 & & & & 0 \\ 1 & 0 & \ddots & & & \\ & \ddots & \ddots & \ddots & & \\ & & \ddots & 0 & \ddots & 1 \\ 0 & & & 1 & & 1 \end{bmatrix}
$$

and $C = \mathrm{diag}(1,0,\dots,0,1)$. Thus when we neglect the term $\varrho(1\text{-}\varrho)C$, the density of u is of the form (3.6), and from Anderson's result, an UMP rejection region against H_1: $\varrho > 0$ is obtained from large values of $\hat{u}'\phi\hat{u}/\hat{u}'\hat{u}$. Since $A = 2I\text{-}2\phi$, this is equivalent to rejecting whenever $d_1 = \hat{u}'A\hat{u}/\hat{u}'\hat{u}$ is too small.

A crucial condition for the approximate optimality of the DW test is that the column space of X be spanned by eigenvectors of ϕ. Much less is known about the power of the DW test when this condition fails. The power of the DW test can even drop to zero for certain regressors. This follows from the fact that for regressions without an intercept, d_1 tends to some constant $\bar{d}$ as $\varrho \to 1$. If $\bar{d}$ is less than the critical level d^{*} corresponding to the given X matrix and α-level, the limiting power of the DW test is 1, otherwise it is zero (neglecting the possibility that $\bar{d} = d^{*}$). Since this result, despite its simplicity, does not seem to be widely known, we will prove it here (see also *Krämer*, 1985b).

The power of the Durbin-Watson test

In what follows, we keep T, X and the ϵ-process fixed and let $\varrho \to 1$. Since the distribution of the DW test statistic does not depend on σ_u^2, we need not worry that σ_u^2 in view of (3.2) tends to infinity as $\varrho \to 1$. One could as well keep σ_u^2 fixed and let $\sigma_\epsilon^2 \to 0$ as $\varrho \to 1$.

Let $i = [1,1,..,1]'$ $(T \times 1)$, and let M denote the column space of X. We assume that i is not an element of M. This is slightly stronger than the requirement that there be no constant in the regression, but will in practice be satisfied for almost all homogeneous regressions.

First we rewrite the test statistic as

$$
\begin{aligned}
d_1 &= \hat{u}'A\hat{u}/\hat{u}'\hat{u} \\
&= u'MAMu/u'Mu \qquad\qquad (3.9) \\
&= u^{*}{}'MAMu^{*}/u^{*}{}'Mu^{*},
\end{aligned}
$$

where

$$
u^{*} = \frac{1}{u_1} u = [1, \frac{u_2}{u_1}, \dots, \frac{u_T}{u_1}].
$$

This holds irrespective of ϱ, which implies

$$\bar{d} = \text{plim}_{\varrho \to 1} \hat{u}\,'A\hat{u}/\hat{u}\,'\hat{u}$$
$$= \text{plim}_{\varrho \to 1} u^{\bullet}\,'MAMu^{\bullet}/u^{\bullet}\,'Mu^{\bullet} \tag{3.10}$$

(given that these probability limits exist). However,

$$u_t^{\bullet} = u_t/u_1$$

$$= \varrho^{t-1} + \Sigma_{i=0}^{t-2}\varrho^i\epsilon_{t-i}/u_1 \tag{3.11}$$

$$= \varrho^{t-1} + (1-\varrho^2)^{1/2}\,\Sigma_{i=0}^{t-2}\varrho^i\epsilon_{t-i}/\tilde{u}_1 \,, \quad (t=2,...,T)$$

where the second equality follows from successively substituting for u_t in (3.1), and where $\tilde{u}_1 = (1-\varrho^2)^{1/2}u_1$.

The critical step is to observe that for any given t,

$$\text{plim}_{\varrho \to 1} u_t^{\bullet} = 1 \,. \tag{3.12}$$

For $t=1$, this is immediate from the definition of $u^{\bullet}$. For $t>1$ it can be seen as follows: The distribution of $\tilde{u}_1$ is $N(0,\sigma_\epsilon^2)$, which does not depend on ϱ. Therefore, the distribution of $1/\tilde{u}_1$ is also fixed, and

$$\Sigma_{i=0}^{t-2}\,\varrho^i\epsilon_{t-i}/\tilde{u}_1 = O_P(1) \tag{3.13}$$

as $\varrho \to 1$. Multiplying this expression by $(1-\varrho^2)^{1/2}$ produces an expression that tends to zero in probability as $\varrho \to 1$. Since in addition $\varrho^t \to 1$ as $\varrho \to 1$, (3.11) now immediately implies that $u^{\bullet} \overset{P}{\to} i$. By assumption, $i \notin M$, so $Mi \neq 0$, and

$$\text{plim}_{\varrho \to 1} u^{\bullet}\,'Mu^{\bullet} = i\,'Mi \neq 0. \tag{3.14}$$

This however implies that

$$\bar{d} = \text{plim}_{\varrho \to 1} d_1 = i\,'MAMi \,/\, i\,'Mi$$

is a finite constant, proving the initial proposition.

A similar result holds when $\varrho \to -1$. Along the same lines as above, it is easily seen that

$$\text{plim}_{\varrho \to -1} u_t^{\bullet} = (-1)^{t+1} \,. \tag{3.15}$$

Denoting $\text{plim}_{\varrho \to -1} u^{\bullet}$ by e, and assuming $e\,'Me \neq 0$, one has

$$\text{plim}_{\varrho \to -1} d_1 = \tilde{d} = e\,'MAMe/e\,'Me \tag{3.16}$$

with some finite constant $\tilde{d}$. Since we now reject H_0 for large values of d_1, the limiting power of the DW test will be one for $\tilde{d} > d^{\bullet}$ and zero otherwise.

Returning to the more relevant case of positive correlation, it is easily seen from the proof of our proposition that the power of the DW test for large values of ϱ depends upon the amount of positive serial correlation in $\hat{i} = Mi$. This is maximized when $\hat{i} = i$, i.e. when i is orthogonal to M, but can be very small otherwise. Since however $\hat{i}$ depends only on X and is therefore known, such situations are easily spotted in practice.

Consider for instance the case $K=1$ and $X_t = 1 + (-1)^t$, i.e. a time series oscillating around a positive constant. Here,

$$\hat{\imath} = [0,1,0,1,...,0,1]' \text{, and (for T even)}$$

$$\bar{d} = i'MAMi/i'Mi = 2(T-1)/T \cong 2 . \tag{3.17}$$

For such series, the null hypothesis of no serial correlation is thus accepted with increasing probability as $\varrho \to 1$.

Figure (3.1) shows the power function for this example, for $T=15$ and $\alpha = 5\%$. The exact critical value d^* for the test as well as the respective rejection probabilities for various ϱ's were computed with *Koerts'* and *Abrahamse's* (1969, p. 159) FQUAD subroutine. In addition, the figure also shows the power function for $K=1$, $T=15$ and X_t equal to a typical exogenous economic time series (taxes in the United States from 1920 to 1934, taken from *Theil* (1971, p. 456). Direct computation shows that $\bar{d} > d^*$ here as well, although the margin is somewhat smaller than for the artificial series. This leads to a higher power for all ϱ and a much steeper decrease toward zero as $\varrho \to 1$.

FIG. 3.1: D-W POWER FUNCTION FOR SELECTED X-SERIES

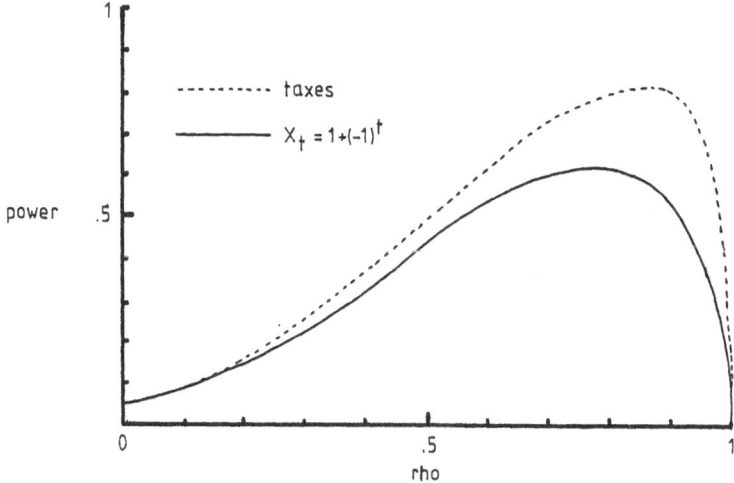

On the other hand, when there is a trend in the data, $\bar{d}$ tends to be rather small and outside the DW acceptance region, which leads to a limiting power of one for the test. For the special case of $K=1$ and a linear trend (i.e. $X_t = t$), it is easily checked that

$\hat{\imath}_t = 1\text{-}3t/(2T+1)$ and

$$\bar{d} = i\,'MAMi/i\,'Mi = 18/T(2T+1)\,, \tag{3.18}$$

which tends to zero as T increases.

Table (3.1) reproduces for further illustration the respective rejection probabilities for various x-series and ϱ-values, plus the corresponding $\bar{d}$ and d^* values. The rejection probabilities in the $\varrho = 1$ column should be viewed as limits as $\varrho \to 1$. The time series appear in descending order of $\bar{d}$ and are (i) $x_t = 1 + (\text{-}1)^t$, (ii) taxes (both as in Figure (3.1)), (iii) government nonwage expenditures (like taxes taken from *Theil* (1971, p. 456), (iv) $x_t = t$, and (v) $x_t = (\text{-}1)^t$. The latter series is an example where (for T even) $\hat{\imath} = i$ and thus $\bar{d} = 0$. Part b of the table gives the analoguous results when a constant is added to the regression. Since both $\bar{d}$ and $\lim_{\varrho \to 1} \Pr(d < d^*)$ cannot be determined here, a question mark appears in these columns.

The table demonstrates that adding an unneccessary constant can quite drastically improve the power of the test that would otherwise drop to zero, contradicting *King* (1981b).

TABLE 3.1: POWER OF THE DURBIN-WATSON TEST FOR VARIOUS X SERIES

(T = 15)

							Pr(d<d*) rho							
	$\bar{d}$	d*	0.0	0.1	0.2	0.3	0.4	0.5	0.6	0.7	0.8	0.9	0.99	0.999	1.0

a) without intercept (K=1)

	$\bar{d}$	d*	0.0	0.1	0.2	0.3	0.4	0.5	0.6	0.7	0.8	0.9	0.99	0.999	1.0
$X_t = 1+(-1)^t$	2.00	1.14	.05	.09	.14	.22	.32	.43	.53	.59	.61	.54	.20	.07	0.00
Taxes	1.62	1.20	.05	.09	.16	.25	.37	.49	.62	.72	.79	.81	.57	.27	0.00
Gov.Exp.	0.67	1.20	.05	.09	.15	.25	.37	.50	.62	.73	.81	.86	.88	.93	1.00
$X_t = t$	0.04	1.20	.05	.09	.16	.25	.37	.50	.63	.75	.83	.90	.96	.98	1.00
$X_t = (-1)^t$	0.02	0.97	.05	.09	.16	.26	.37	.54	.70	.82	.91	.96	.99	1.00	1.00

b) with intercept (K=2)

	$\bar{d}$	d*	0.0	0.1	0.2	0.3	0.4	0.5	0.6	0.7	0.8	0.9	0.99	0.999	1.0
$X_t = 1+(-1)^t$	?	1.09	.05	.09	.14	.23	.34	.46	.59	.70	.79	.86	.90	.90	?
Taxes	?	1.22	.05	.09	.14	.22	.32	.44	.55	.66	.76	.83	.87	.88	?
Gov.Exp.	?	1.31	.05	.09	.14	.22	.31	.41	.51	.66	.68	.75	.80	.81	?
$X_t = t$	?	1.36	.05	.09	.14	.22	.32	.42	.52	.61	.68	.72	.74	.74	?
$X_t = (-1)^t$	?	1.09	.05	.09	.14	.23	.34	.41	.59	.70	.79	.86	.90	.90	?

Some readers may find the notion disturbing that a statistical procedure can be improved upon by adding something to a model one knows is incorrect. This apparent contradiction is resolved by noting that in such cases, DW is not the optimal test (see *King*, 1985), as OLS is not the optimal estimator. In fact it is easily seen that OLS coefficient estimates can likewise be improved by adding an unnecessary constant: Consider the case $K=1$, and let X be such that $X'i = \Sigma_{t=1}^{T} x_t \neq 0$. The variance of the OLS coefficient estimate is then

$$\text{var}(\hat{\beta}) = q_u^2 X'VX/(X'X)^2 . \tag{3.19}$$

Adding an unnecessary constant to the regression produces an estimate $\tilde{\beta} = \dot{X}'y/\dot{X}'\dot{X}$, where $\dot{X}_t = X_t - \overline{X}$. From $\tilde{\beta} - \beta = \dot{X}'U/\dot{X}'\dot{X}$, the variance of this estimate is

$$\text{var}(\tilde{\beta}) = \sigma_u^2 \, \dot{X}'V\dot{X}/(\dot{X}'\dot{X})^2 . \tag{3.20}$$

From (3.3), $\lim_{\varrho \to 1} V = \overline{V} = ii'$, and therefore $\dot{X}'V\dot{X} \to 0$ as $\varrho \to 1$. On the other hand, $X'i \neq 0$, and thus $X'VX \to C > 0$ as $\varrho \to 1$. Together, this implies that

$$\text{var}(\tilde{\beta})/\text{var}(\hat{\beta}) \to 0 \quad \text{as} \quad \varrho \to 1, \tag{3.21}$$

i.e. the inclusion of an unnecessary constant greatly improves the efficiency of the OLS coefficient estimate.

Tillman (1975) and *King* (1985) have also pointed out that the power of the DW test might drop to zero, without however proving this. The fact that this rather distressing property is not better known might be due to the prevalence of inhomogeneous regressions, both in theory and practice. In addition, all previous power investigations that we know of (*Koerts* and *Abrahamse* (1969), *Abrahamse* and *Louter* (1971), *Habibaghi* and *Pratschke* (1972), *Dubbelman* (1972), *Phillips* and *Harvey* (1974), *L'Esperance* and *Taylor* (1975), *Schmidt* and *Guilkey* (1975), *Smith* (1977), *Dubbelman* et al. (1978), *Dent* and *Styan* (1978), or *King* (1985)) only consider ϱ's less than or equal to $\varrho = 0.9$. From figure (3.1), it is however obvious that trouble with the power of the test, if any, might well begin only beyond $\varrho = 0.9$.

The DW inconclusive region

A major problem with the DW test used to be that the rejection region depends not only on the significance level α of the test, but also on the regressor matrix X. *Durbin* and *Watson* (1951) gave the familiar bounds d_L and d_u for d that depend only on α, T and K, such that (when testing against positive serial correlation) H_0 is rejected when $d_l < d_L$ and accepted when $d_l > d_u$. No conclusion is possible from their tables when $d_L \leq d_l \leq d_u$.

Kramer (1971) and *Farebrother* (1980) have extended the initial DW tables, which are valid only for inhomogeneous regressions, to regressions without an intercept, and extensions to extreme sample sizes are provided by *Savin* and *White* (1977). There have also been efforts to narrow the inconclusive region by putting restrictions on X (*King*, 1981b; *Bartels* et al., 1982; *King*, 1983b) or to approximate the true critical value d^* (see *Durbin* and *Watson*, 1971; *Harrison*, 1972; *Maddala*, 1977, p. 285; or *King*, 1983a for convenient surveys of this literature). Most of this work however appears to be obsolete, since the exact DW critical values d* can be computed by easily available computer programs such as *Koerts* and *Abrahamse*'s (1969) FQUAD subroutine and are also provided by standard software packages. The IAS-SYSTEM for instance routinely computes the prob-value of the sample, i.e. $Pr(d_1 \leq d_1')$, where d_1' is the sample outcome. If this is less than α, the null hypothesis of no serial correlation is rejected, otherwise it is accepted.

The availability of computer procedures to get rid of the inconclusive region of the DW test has also made obsolete various attempts to overcome this problem by basing the test statistic on residuals other than the OLS-residuals û. Most prominent among these are *Theil's* (1965) BLUS residuals, or recursive residuals. A convenient survey of this literature may again be found in *King* (1983 a). Autocorrelation tests based on such residuals have often proved less powerful than the exact DW test, and we will not follow this line any further. For tests against different types of alternatives, such residuals, in particular recursive residuals, do have some merits, and we will make heavy use of them later.

Relationship to the LM principle

Let in general the disturbance vector u in the linear regression model have covariance matrix $cov(u) = \sigma_u^2 V$, where V is a matrix valued function with known functional form, of some unknown parameter θ, and $V(\theta = 0) = I$. Both σ^2 and θ are unknown. This comprises first order autoregressive errors as a special case, where $\theta = \varrho$ and V is as in (3.3), but permits also heteroskedastic disturbances.

The log likelihood function of the model (2.2) is then

$$L(\beta, \sigma^2, \theta) = -\frac{T}{2} \log(2\pi\sigma_u^2) - \frac{1}{2} \log |V(\theta)| - \frac{1}{2} (y - X\beta)' V(\theta)^{-1} (y - X\beta)/\sigma_u^2 \qquad (3.22)$$

The ML estimates for β, σ^2 and θ are obtained by solving

$$\frac{\partial L}{\partial \beta} = -[(X'V(\theta)^{-1}X)\beta - X'V(\theta)^{-1}y]/\sigma_u^2 = 0 , \qquad (3.23)$$

$$\frac{\partial L}{\partial \sigma^2} = -\frac{T}{2\sigma_u^2} + \frac{1}{2}(y - X\beta)'V(\theta)^{-1}(y - X\beta)/\sigma_u^4 = 0 , \qquad (3.24)$$

$$\frac{\partial L}{\partial \theta} = d(\theta) - \frac{1}{2}(y - X\beta)'A(\theta)(y - X\beta)/\sigma_u^2 = 0 , \qquad (3.25)$$

where

$$d(\theta) = -\frac{1}{2}|V(\theta)^{-1}|\frac{\partial |V(\theta)|}{\partial \theta} \quad \text{and} \quad A(\theta) = \partial V(\theta)^{-1}/\partial \theta.$$

Now we want to test H_0: $\theta = 0$ against H_1: $\theta \neq 0$. The LM approach requires the restricted ML estimates for θ, β and σ^2 . These are $\hat{\theta}=0$, $\hat{\beta}=(X\,'X)^{-1}X\,'y$ and $\hat{\sigma}^2 = \hat{u}\,'\hat{u}/T$, where $\hat{u}=Mu$ as in (2.9). The LM test of H_0: $\theta=0$ thus rejects the null hypothesis for large absolute values of

$$\frac{\partial L}{\partial \theta}\,(\theta=0) = d(0) - \hat{u}\,'A(0)\hat{u}/2\hat{\sigma}^2_u. \tag{3.26}$$

For the special case where $\theta=\varrho$ and V is as in (3.3), i.e. where the alternative is that the disturbances follow an AR(1) process, we have $d(0)=0$ (to see this requires some manipulation of $|V(\varrho)|$, which we omit here),

$$V(\varrho)^{-1} = (1-\varrho^2)^{-1}\begin{bmatrix} 1 & -\varrho & & & 0 \\ -\varrho & 1+\varrho^2 & \ddots & & \\ & \ddots & \ddots & \ddots & \\ & & \ddots & 1+\varrho^2 & -\varrho \\ 0 & & & -\varrho & 1 \end{bmatrix} \tag{3.27}$$

and

$$A(\varrho) = \frac{\partial V(\varrho)^{-1}}{\partial \varrho} = (1-\varrho^2)^{-2}\begin{bmatrix} 2\varrho & -\varrho^2-1 & & & 0 \\ -\varrho^2-1 & 4\varrho & \ddots & & \\ & \ddots & \ddots & \ddots & \\ & & \ddots & 4\varrho & -\varrho^2-1 \\ 0 & & & -\varrho^2-1 & 2\varrho \end{bmatrix}, \tag{3.28}$$

i.e.

$$A(0) = \begin{bmatrix} 0 & -1 & & & 0 \\ -1 & 0 & \ddots & & \\ & \ddots & \ddots & \ddots & \\ & & \ddots & 0 & -1 \\ 0 & & & -1 & 0 \end{bmatrix}. \tag{3.29}$$

This implies that (3.26) reduces to

$$-\frac{T}{2}\,\hat{u}\,'A(0)\hat{u}/\hat{u}\,'\hat{u} = T\hat{\varrho}\,, \tag{3.30}$$

where

$$\hat{\varrho} = \Sigma^T_{t=2} \hat{u}_t \hat{u}_{t-1} / \Sigma^T_{t=1} \hat{u}_t^2 \qquad (3.31)$$

is the familiar estimate for ϱ. The LM test therefore rejects H_0: $\varrho = 0$ in favour of H_1: $\varrho \neq 0$ whenever $|\hat{\varrho}|$ is too large. The appropriate rejection region follows from the limiting $N(0,1)$ distribution of $T^{1/2}\hat{\varrho}$, i.e. the LM approach yields only an asymptotic test.

The same approach also produces tests against H_1: $u_t = \varrho_i u_{t-i} + \epsilon_t$, where $i > 1$. The test statistic is then $T^{1/2}\hat{\varrho}_i$, where $\hat{\varrho}_i$ is the i'th order empirical autocorrelation coefficient of the OLS residuals.

This procedure relates to the DW test as follows: from (3.4), the DW test statistic can be written as

$$d_1 = \Sigma^T_{t=2} (\hat{u}_t - \hat{u}_{t-1})^2 / \hat{u}'\hat{u}$$

$$= (\Sigma^T_{t=2} \hat{u}_t^2 + \Sigma^{T-1}_{t=1} \hat{u}_t^2 - 2\Sigma^T_{t=2} \hat{u}_t \hat{u}_{t-1}) / \hat{u}'\hat{u} \qquad (3.32)$$

$$= 2 - 2\hat{\varrho} - \hat{u}_1^2 / \hat{u}'\hat{u} - \hat{u}_T^2 / \hat{u}'\hat{u} .$$

Since $T^{1/2}\hat{u}_t^2 / \hat{u}'\hat{u} \xrightarrow{P} 0$ for any given t as $T \to \infty$, we have

$$\text{plim} \, [\, T^{1/2}\hat{\varrho} - T^{1/2}\cdot\frac{1}{2}(2 - d_1) \,] = 0 , \qquad (3.33)$$

i.e. both quantities have the same limiting distribution. Rejecting H_0 for large absolute values of $\hat{\varrho}$ is thus asymptotically equivalent to rejecting for d_1 far away from 2.

Dynamic regressions

Both the DW test and the LM procedure based on $\hat{\varrho}$ are misleading when there are lagged dependent variables among the regressors, since the nominal α-level of the test is not even attained asymptotically (*Nerlove* and *Wallis*, 1966). *Durbin* (1970) however shows that

$$h = (T/(1 - T\hat{V}(\hat{\alpha})))^{1/2}\hat{\varrho} \qquad (3.34)$$

is again asymptotically standard normal under H_0, where $\hat{V}(\hat{\alpha})$ is the OLS-based estimate of the variance of the coefficient of y_{t-1}. This Durbin h-test applies also when X contains higher order lags of y in addition to lag 1, and can likewise be derived from the LM principle (*Aldrich*, 1978; *Breusch*, 1978; *Godfrey*, 1978b). For the special case where

$$y_t = \alpha y_{t-1} + \beta x_t + u_t , \qquad (3.35)$$

and the u_t's are as in (3.1), this can be seen as follows:

First transform the data so that

$$y_t^* = y_t - \varrho y_{t-1}, \, x_t^* = x_t - \varrho x_{t-1} \, (t = 1,...,T). \qquad (3.36)$$

Here we take the pre-sample values y_0 and x_0 as known and fixed constants. One could as well discard the first two observations without affecting the results. The transformed model is then

$$y_t^* = \alpha y_{t-1}^* + \beta x_t^* + \epsilon_t , \tag{3.37}$$

where ϵ_t is independent of the contemporaneous and previous regressor values. The log likelihood function is therefore

$$L(y^*, \alpha, \beta, \varrho, \sigma_\epsilon^2) = -\frac{T}{2}(\log 2\pi\sigma_\epsilon^2) \tag{3.38}$$

$$-\frac{1}{2\sigma_\epsilon^2}\{\Sigma_{t=1}^T[(y_t - \alpha y_{t-1} - \beta x_t) - \varrho(y_{t-1} - \alpha y_{t-2} - \beta x_{t-1})]^2\}$$

$$= -\frac{T}{2}(\log 2\pi\sigma_\epsilon^2) - \frac{1}{2\sigma_\epsilon^2}\Sigma_{t=1}^T \epsilon_t^2 .$$

Denoting the log likelihood of the t'th observation by l_t as in (2.42), we have

$$\frac{\partial l_t}{\partial \alpha} = \frac{1}{\sigma_\epsilon^2}(y_{t-1} - \varrho y_{t-2})\epsilon_t , \tag{3.39}$$

$$\frac{\partial l_t}{\partial \beta} = \frac{1}{\sigma_\epsilon^2}(x_t - \varrho x_{t-1})\epsilon_t , \text{ and} \tag{3.40}$$

$$\frac{\partial l_t}{\partial \varrho} = \frac{1}{\sigma_\epsilon^2}(y_{t-1} - \alpha y_{t-2} - \beta x_{t-1})\epsilon_t . \tag{3.41}$$

Under H_0, $\varrho = 0$, and (3.41) reduces to

$$\frac{\partial l_t}{\partial \varrho} = \frac{1}{\sigma_u^2} u_t u_{t-1} \tag{3.42}$$

If we evaluate this at the restricted ML estimate $\bar{\theta} = (0, \hat{\alpha}, \hat{\beta}, \hat{\sigma}_u^2)'$, where $\hat{\alpha}, \hat{\beta}$ and $\hat{\sigma}_u^2 = \frac{1}{T}\Sigma \hat{u}_t^2$ are simply the OLS estimates, we obtain (setting $\hat{u}_0 = 0$)

$$\frac{\partial L}{\partial \varrho} = \frac{1}{\sigma_u^2}\Sigma_{t=1}^T \hat{u}_t \hat{u}_{t-1} = T\hat{\varrho} , \tag{3.43}$$

which is identical to the expression (3.30) in the non-dynamic model. In both cases, we reject H_0 for large absolute values of $\hat{\varrho}$. The crucial difference is that in the dynamic case, $\hat{\varrho}$ is more concentrated around $\varrho = 0$. This implies that taking $\hat{\varrho}$ to have a variance of $1/T$ as in the standard model will result in a test in which the actual size is smaller than the nominal size, even asymptotically.

The correct normalizing factor for $\hat{\varrho}$ is found by computing the information matrix. Since this is easily seen to be block diagonal with respect to σ_ϵ^2 and the other parameters, we can disregard the part of the information matrix relating to σ_ϵ^2. The rest is estimated using (2.43), where (3.39), (3.40) and (3.41) are evaluated at the restricted ML estimates $(0, \hat{\alpha}, \hat{\beta}, \hat{\sigma}_u^2)$. This yields

$$\hat{I}(\theta) = \Sigma_{t=1}^T \left(\frac{\partial l_t}{\partial \theta}(\bar{\theta})\right)\left(\frac{\partial l_t}{\partial \theta}(\bar{\theta})\right)'$$

$$
= \left[\frac{1}{\sigma_u^2} \begin{bmatrix} \Sigma_{t=1}^T y_{t-1}^2 & \Sigma_{t=1}^T y_{t-1} x_t & T \\ \Sigma_{t=1}^T y_{t-1} x_t & \Sigma_{t=1}^T x_t^2 & 0 \\ T & 0 & T \end{bmatrix} \right].
\tag{3.44}
$$

The upper right and lower left elements of this matrix are for instance obtained as

$$
E \left[\Sigma_{t=1}^T \frac{\partial l}{\partial \alpha} \frac{\partial l}{\partial \varrho} (0, \hat{\alpha}, \hat{\beta}, \hat{\sigma}_u^2) \right]
$$

$$
= E \left[\Sigma_{t=1}^T \frac{1}{\sigma_u^2} y_{t-1} u_t \frac{1}{\sigma_u^2} u_{t-1} u_t \right]
$$

$$
= \frac{1}{(\sigma_u^2)^2} \Sigma_{t=1}^T E(u_t^2 u_{t-1}^2) = T.
\tag{3.45}
$$

These elements are zero in the standard case. Here, they produce

$$
[T(1 - T \, \hat{V}(\hat{\alpha}))]^{-1}
\tag{3.46}
$$

as the bottom right element of $I(\hat{\theta})^{-1}$. Since $\partial L/\partial \alpha(\bar{\theta}) = \partial L/\partial \beta(\bar{\theta}) = 0$ under H_0, (3.46) is at the same time the normalizing factor for $\partial L/\partial \varrho$. The LM statistic is therefore

$$
\xi_{LM} = S(\bar{\theta})' I(\bar{\theta})^{-1} S(\bar{\theta})
$$

$$
= [T/(1 - T\hat{V}(\hat{\alpha}))]\hat{\varrho}^2
\tag{3.47}
$$

which is the square of (3.34). This shows that the (two-sided) Durbin h test is also the LM test for H_0: $\varrho = 0$.

Along similar lines, *Godfrey* (1978) has derived a test against higher order autoregressive and moving average disturbances. The alternatives are either

$$
u_t = \varrho_1 u_{t-1} + \ldots + \varrho_q u_{t-q} + \epsilon_t
\tag{3.48}
$$

(AR(q)), or

$$
u_t = \epsilon_t + \varrho_1 \epsilon_{t-1} + \ldots + \varrho_q \epsilon_{t-q}
\tag{3.49}
$$

(MA(q)), where the ϵ_t's are NID$(0, \sigma_\epsilon^2)$. The test statistic against both alternatives is

$$
1 = T\hat{u}' \hat{U}[\hat{U}' \hat{U} - \hat{U}' X (X'X)^{-1} X' \hat{U}] \hat{U}' \hat{u}/\hat{u}' \hat{u},
\tag{3.50}
$$

where $\hat{U} = [\hat{u}^{(1)}, \ldots, \hat{u}^{(q)}]$ and

$$
\hat{u}^{(i)} = [0, \ldots, 0, \hat{u}_1, \ldots, \hat{u}_{T-i}]' \quad (i = 1, \ldots, q).
$$

The test statistic equals TR^2 in a regression of $\hat{u}$ against $[\hat{U}, X]$, and has an asymptotic $\chi_{(q)}^2$ distribution under the null hypothesis of no correlation. Significantly large values again imply that the sample data are inconsistent with H_0. This holds whether or not there are lagged dependent variables in the regression, but does not generalize to mixed autoregressive moving average (ARMA) processes.

Simultaneous equations

A Durbin-Watson type test for autocorrelation of the disturbances can also be formed in the simultaneous equation context (see *Harvey* and *Phillips*, 1980). To this purpose, consider the standard linear simultaneous equation system

$$YB + Z\Gamma = U , \tag{3.51}$$

where Y is a $T \times G$ matrix of observations on G endogenous variables, Z is a $T \times K$ matrix of observations on K exogenous variables, and U is a $T \times G$ matrix of structural disturbances, with rows $NID(0,\Sigma)$ under H_0. B and Γ are respectively $G \times G$ and $K \times G$ matrices of structural coefficients.

Write the equation of interest as

$$y_1 = Y_2\beta + Z_1\gamma + u_1 , \tag{3.52}$$

where y_1 and Y_2 are, respectively, $T \times 1$ and $T \times g_1$ matrices of observations on $g_1 + 1$ endogenous variables and Z_1 is a $T \times k_1$ matrix of observations on k_1 exogenous variables with full column rank. The aim is to derive a test for the serial independence of the components of u_1.

To this purpose, let $Y_1 = [y_1:Y_2]$, with reduced form $Y_1 = Z\Pi_1 + V_1$, and partition $\Pi_1 = [\pi_1:\Pi_2]$ and $V_1 = [v_1:V_2]$ conformably. The reduced form for y_1 is then

$$y_1 = Z\pi_1 + v_1, \tag{3.53}$$

and (3.52) can also be written as

$$y_1 = Z\Pi_2\beta + Z_1\gamma + V_2\beta + u_1 . \tag{3.54}$$

Equating the stochastic components of (3.53) and (3.54) shows that

$$u_1 = v_1 - V_2\beta = V_1\beta_0, \tag{3.55}$$

where $\beta_0 = [1,-\beta']'$. The basic idea now is to approximate the unobservable disturbance vector u_1 by estimating V_1 and β_0 separately.

Let

$$\beta^* = (Y_2'(P_z - P_{z_1})Y_2)^{-1}Y_2'(P_z - P_{z_1}) y_1 \tag{3.56}$$

be the 2SLS-estimate for β, where $P_z = Z(Z'Z)^{-1}Z'$ and $P_{z_1} = Z_1(Z_1'Z_1)^{-1}Z_1'$ are the orthogonal projections on the column spaces of Z and Z_1, respectively. It is immediately verified from (3.56) that β^* depends on V only through $(P_z - P_{z_1})V_1$, and is thus independent of $\hat{V}_1 = (I-P_z)V_1$ (since $(I-P_z)(P_z - P_{z_1}) = 0$). We can therefore approximate u_1 by

$$u_1^* = \hat{V}_1\beta_0^* , \tag{3.57}$$

where $\hat{V}_1$ is independent of β_0^* (u_1^* can also be obtained as the residual vector from regressing $y_1 - Y_2\beta^*$ on Z). The conditional distribution of u_1^*, given β_0^*, is thus $N(0,\omega^{*2}(I-P_z))$, where $\omega^{*2} = \beta_0^*{}'\Omega\beta_0^*$ and Ω is the covariance matrix of V_1. Conditional on β_0^*, u_1^* therefore has the same distribution as an OLS residual vector from a regression on Z, and may thus be used as input to the standard DW test statistic d_1 as defined in (3.4). Since in addition ω^{*2} cancels out when evaluating d_1, this also holds unconditionally and

Since in addition ω^{*2} cancels out when evaluating d_1, this also holds unconditionally and provides the desired test.

It should be warned however that this procedure is only valid when there are no lagged endogenous variables among the exogenous (predetermined) variables of the system (3.51), and tests for autocorrelation among the components of any column of U, not only among the elements of u_1.

Other alternatives to the Durbin-Watson test

Since the DW procedure has been specifically designed to test for AR(1) disturbances, it might have poor power when the disturbances are generated by a different scheme, even with nonstochastic regressors and ruling out the perverse limiting cases discussed earlier. *Wallis* (1972) for instance has suggested that an AR(4) disturbance process $u_t = \varrho u_{t-4} + \epsilon_t$ might be a more reasonable alternative with quarterly data, which on the analogy to the DW test leads to the test statistic

$$d_4 = \Sigma_{t=5}^{T}(\hat{u}_t - \hat{u}_{t-4})^2 / \Sigma_{t=1}^{T} \hat{u}_t^2. \tag{3.58}$$

Similar to DW, the procedure based on d_4 is a bounds test, with analoguous rules to accept or to reject the null hypothesis. *Wallis* (1972) provides some tables with bounds for selected values of T, K and α. Given X and α, the exact critical value could however in principle be computed by the same methods used above.

For testing H_0: $\varrho_1 = \varrho_2 = 0$, when the alternative is H_1: $u_t = \varrho_1 u_{t-1} + \varrho_2 u_{t-2} + \epsilon_t$, *Schmidt* (1972) has also suggested to use $d_1 + d_2$.

Godfrey and *Tremayne* (1978) have extended the Wallis test to dynamic regressions, suggesting the test statistic

$$h_4 = (1 - \frac{1}{2} d_4)(T/(1 - T\tau'\hat{V}(\beta)\tau))^{1/2}, \tag{3.59}$$

where d_4 is from (3.58) and $\hat{V}(\beta) = \hat{\sigma}^2 (X'X)^{-1}$ is the covariance matrix of $\hat{\beta}$ as estimated by OLS. τ is a vector depending on the maximum lag m of the dependent variable. For

$$m = 1 : \tau' = (\hat{\beta}_1^2, 0, \ldots, 0)$$
$$m = 2 : \tau' = (\hat{\beta}_1^3 + 2\hat{\beta}_1\hat{\beta}_2\hat{\beta}_1^2 + \hat{\beta}_2, 0, \ldots, 0)$$
$$m = 3 : \tau' = (\hat{\beta}_1^3 + 2\hat{\beta}_1\hat{\beta}_2 + \hat{\beta}_3\hat{\beta}_1^2 + \hat{\beta}_2\hat{\beta}_1, 0, \ldots, 0)$$
$$m > 3 : \tau' = (\hat{\beta}_1^3 + 2\hat{\beta}_1\hat{\beta}_2 + \hat{\beta}_3\hat{\beta}_1^2 + \hat{\beta}_2\hat{\beta}_1, 1, 0, \ldots, 0).$$

h_4 is asymptotically N(0,1) under H_0. Note that both Durbin's h and h_4 cannot be computed for $T\hat{V}(\beta_1)$ or $T\tau'\hat{V}(\beta)\tau$ greater than unity. This reminds us that asymptotic results are not without problems in finite samples.

For disturbances that are AR(1) but not necessarily stationary, (i.e. where $|\varrho| \gtrless 1$ in 3.1), *Berenblut* and *Webb* (1973) recommend the statistic

$$g = \bar{u}'B\bar{u} / \hat{u}'\hat{u}, \tag{3.60}$$

where $\hat{u}$ is the residual vector from a regression of the first differences of y on the first differences of X without a constant term and B equals the differencing matrix A from (3.5) with the only exception that 2 rather than 1 appears in the upper left corner. Provided that the original equation contains a constant term, g and the familiar Durbin-Watson bounds can be used to test for serial correlation in the same way as the Durbin-Watson d_1 is used. For large ϱ, this procedure is in general more powerful than the DW test, which is not surprising in view of our results above.

Similar to Berenblut and Webb, *King* (1982) has proposed the statistic

$$S(\bar{\varrho}) = \dot{u}\,'V(\bar{\varrho})^{-1}\dot{u}/\hat{u}\,'\hat{u}\,, \qquad (3.61)$$

where $\dot{u}$ is the Generalized Least Squares (GLS) residual vector assuming (3.3) and $\varrho = \bar{\varrho}$, and where $V(\bar{\varrho})$ is as in (3.3). Again this test rejects H_0 in favour of $H_1 : \varrho > 0$ for small values of the test statistic, and is superior to the DW test in the neighborhood of $\varrho = \bar{\varrho}$.

For a more detailed discussion of the pros and cons of these and various other approaches which we have not surveyed here, we refer the reader to *Judge* et al. (1980, section 5.4) or *King* (1983 a).

b) Heteroskedasticity

There are probably more tests for heteroskedasticity, i.e. nonidentical variance of the disturbances u_t in the regression equation (2.1), then for anything else. None of these procedures however has achieved the wide-spread acceptance that the Durbin-Watson test enjoys when testing for autocorrelation.

Comparing empirical variances

The most popular - and simple - test is due to *Goldfeld* and *Quandt* (1965). It depends on one's ability to rank the observations in the regression model (2.1) according to increasing variance, given that the disturbances are heteroskedastic. Familiar criteria for ordering the sample are (i) time or (ii) the absolute value of some independent variable of the model. The sample is then split into $y_1 = X_1\beta + u_1$ and $y_2 = X_2\beta + u_2$, where $y_1(T_1 \times 1)$ contains the observations with the smaller variance and $y_2((T-T_1) \times 1)$ contains the observations with the larger variances. The test statistic is

$$R = \frac{\hat{u}_2\,'\hat{u}_2\,/\,(T-T_1-K)}{\hat{u}_1\,'\hat{u}_1\,/\,(T_1-K)}\,, \qquad (3.62)$$

where $\hat{u}_i$ is the OLS residual vector from the i'th regression (i = 1,2). Under H_0, R has an F-distribution with $T-T_1-K$ and T_1-K degrees of freedom, and we reject H_0 whenever R is too large.

Due to the form of the test statistic (3.62), the Goldfeld-Quandt test is also known as the Variance Ratio test. Obviously, it can be applied only if $K < T_1 < T-K$, and for $r(X_1) = r(X_2) = K$, which we will assume throughout.

It is easily seen that the test statistic (3.62) retains its null distribution even when the regression coefficients β differ across subsamples. This is a trivial but important property of the Goldfeld-Quandt test, which follows from the invariance of the residual vectors $\hat{u}_i$ to a shift in β. The implicit null hypothesis of the Goldfeld-Quandt test is therefore H_0^*: $\sigma_1^2 = \sigma_2^2$, $\beta_1 \neq \beta_2$ (possibly), where σ_i^2 and β_i are respectively the variance and the regression coefficients in the i'th sample $(i = 1,2)$.

Moreover, the Goldfeld-Quandt test is equivalent to the Likelihood Ratio (LR) test if the null hypothesis is put like this. To see this, let $T_2 = T-T_1$. The log likelihood function of the i'th subsample is then (compare 2.27)

$$L(\beta_i, \sigma_i^2, y_i) = -\frac{T_i}{2}\ln(2\pi) - \frac{T_i}{2}\ln\sigma_i^2 - \frac{(y_i - X_i\beta_i)'(y_i - X_i\beta_i)}{2\sigma_i^2}$$

which leads to the familiar OLS-based Maximum Likelihood (ML) estimates $\hat{\beta}_i = (X_i'X_i)^{-1}X_i'y_i$ and $\hat{\sigma}_i^2 = \hat{u}_i'\hat{u}_i/T_i$ $(i = 1,2)$. The maximized values of the log likelihood functions are therefore under the alternative

$$l_i = -\frac{T_i}{2}\ln(2\pi) - \frac{T_i}{2} - \frac{T_i}{2}\ln(\hat{\sigma}_i^2) .$$

Under H_0^*, the estimates of coefficients and residuals are exactly the same as under the alternative, but the variance estimate is

$$\tilde{\sigma}^2 = (\hat{u}_1'\hat{u}_1 + \hat{u}_2'\hat{u}_2)/T .$$

This yields the following maximum for the log likelihood function:

$$l = -\frac{T}{2}\ln(2\pi) - \frac{T}{2} - \frac{T}{2}\ln(\tilde{\sigma}^2) .$$

The test statistic for the LR test is therefore

$$LR = 2(l - l_1 - l_2)$$
$$= T\ln(\tilde{\sigma}^2) - T_1\ln(\hat{\sigma}_1^2) - T_2\ln(\hat{\sigma}_2^2) ,$$

a monotonic transformation of (3.62). LR is asymptotically χ_1^2 under H_0, and would imply the same rejection region as the test statistic (3.62) if its exact finite sample distribution were known.

The power of the Goldfeld-Quandt test may be increased by omitting some r central observations before computing $\hat{\sigma}_1^2$ and $\hat{\sigma}_2^2$, so that under heteroskedasticity, residuals in the numerator correspond to relatively large variances and residuals in the denominator correspond to relatively small variances. There has been some debate in the literature about the optimum value of r, but so far no unanimous verdict has been reached. *Goldfeld* and *Quandt* (1965) suggest $r = 8$ for $T = 30$ and $r = 16$ for $T = 60$.

A related test by *Harrison* and *McCabe* (1979) is based solely on OLS-residuals from the overall regression. The test statistic is simply

$$b = \hat{u}_1'\hat{u}_1/\hat{u}'\hat{u}, \tag{3.63}$$

where $\hat{u}_1$ is some subset of the OLS residuals $\hat{u}$, such that under the alternative the variance is smallest. The dimension T_1 of the subvector $\hat{u}_1$ is arbitrary, except if there is additional knowledge about the form of heteroskedasticity. This is similar to the choice of r for the Goldfeld-Quandt test.

Under H_0, the statistic b should be close to T_1/T, and we reject H_0 whenever b is too small. The problem is that the distribution of b is much harder to evaluate than that of the Goldfeld-Quandt statistic R, which explains why the Harrison-McCabe test is so rarely used in practice. Harrison and McCabe show that b is bounded by random variables b_L and b_u that depend only on T, T_1 and K, and suggest a bounds test similar to the Durbin Watson test.

The Harrison-McCabe test can be derived from the Lagrange Multiplier principle, if the alternative is suitably defined. To see this, assume the sample is ordered such that $E(u_t^2) = \sigma_1^2$ for $t \leq T_1$, and $E(u_t^2) = \sigma_2^2$ for $t > T_1$. Assume in addition that T_1 is known. Let $\theta = (\sigma_2^2 - \sigma_1^2)/\sigma_2^2$. The disturbance covariance matrix is then of the form

$$\sigma_2^2 V(\theta) = \sigma_2^2 \text{diag}(1-\theta,...,1-\theta,1,...,1),$$

with H_0: $\theta = 0$ and H_1: $\theta > 0$. This corresponds to the situation from p. 25, so the former approach applies here as well.

We have $|V(\theta)| = (1-\theta)^{T_1}$, i.e.

$$d(\theta) = -2^{-1}|V(\theta)|^{-1} \partial|V(\theta)|/\partial\theta$$
$$= 2^{-1}(1-\theta)^{-T_1} T_1(1-\theta)^{T_1-1} \tag{3.64}$$

Moreover,

$$V(\theta)^{-1} = \text{diag}((1-\theta)^{-1},...,(1-\theta)^{-1},1,..,1), \text{ i.e.}$$
$$A(\theta) = \partial V(\theta)^{-1}/\partial\theta$$
$$= \text{diag}((1-\theta)^{-2},...,(1-\theta)^{-2},0,...,0). \tag{3.65}$$

The partial derivative of the log likelihood function with respect to θ, evaluated at the restricted Maximum Likelihood estimates $\theta = 0$, $\beta = \hat{\beta}$, $\sigma^2 = \hat{\sigma}^2$ where $\hat{\beta}$ and $\hat{\sigma}^2$ are again the familiar OLS estimates, is therefore in view of (3.26) given by

$$\frac{\partial L}{\partial \theta} = d(0) - \hat{u}'A(0)\hat{u}/2\hat{\sigma}^2$$
$$= \frac{T_1}{2} - \frac{T}{2}\hat{u}_1'\hat{u}_1/\hat{u}'\hat{u} \tag{3.66}$$
$$= \frac{T_1}{2}(1 - \frac{\hat{u}_1'\hat{u}_1/T_1}{\hat{u}'\hat{u}/T}).$$

The term in parentheses in the last expression should be close to zero when H_0 is true, so we reject H_0 whenever (3.66) is positive and too large, which is equivalent to rejecting whenever the Harrison-McCabe b statistic is too small.

A general LM test

Assume that the disturbance variances are given by

$$E(u_t^2) = h(z_t'\alpha) \quad (t = 1,...,T),$$ (3.67)

where h is some twice differentiable function and z_t is a px1 vector of nonstochastic variables (the first of which is unity) which may or may not be equal to the independent variables of the regression equation under test. In practice, one often uses products and cross products of the independent variables.

Given (3.67), the null hypothesis of homoskedastic disturbances is obviously equivalent to H_0: $\alpha_2 = ... = \alpha_p = 0$. *Breusch* and *Pagan* (1979) suggest a simple Lagrange Multiplier test to determine whether or not this is true.

The log-likelihood function of the regression $y_t = x_t'\beta + u_t$ is

$$l(\beta,\alpha) = -\frac{T}{2} \log(2\pi) - \frac{1}{2} \Sigma_{t=1}^T \log \sigma_t^2 - \frac{1}{2} \Sigma_{t=1}^T (y_t - x_t'\beta)^2/\sigma_t^2,$$ (3.68)

where $\sigma_t^2 = h(z_t'\alpha)$. The derivative of the log likelihood with respect to α is therefore

$$\frac{\partial L}{\partial \alpha}(\beta,\sigma^2) = -\frac{1}{2}\Sigma h'(z_t'\alpha)z_t/\sigma_t^2 + \frac{1}{2}\Sigma(y_t - x_t'\beta)^2 h'(z_t'\alpha) z_t/\sigma_t^4$$

$$= \frac{1}{2}\Sigma h'(s_t)z_t(u_t^2\sigma_t^{-4} - \sigma_t^{-2}),$$ (3.69)

where $s_t = z_t'\alpha$ and $h'(s_t) = \partial h(s_t)/\partial s_t$. Under H_0, $\sigma_t^2 = h(\alpha_1) = \sigma^2$ does not depend on t, and the restricted ML estimates for the parameters in the model are again the familiar OLS estimates $\hat{\beta}$ and $\hat{\sigma}^2$. Evaluating (3.69) at these quantities yields

$$\frac{\partial L}{\partial \alpha}(\beta,\hat{\sigma}^2) = \frac{1}{2}\hat{\sigma}^{-2}h'(\hat{\alpha}_1) \Sigma z_t(\hat{u}_t^2\hat{\sigma}^{-2}-1),$$ (3.70)

where $\hat{\alpha}_1 = h^{-1}(\hat{\sigma}^2)$.

Consider next the information matrix that we need to calculate the LM statistic. Since this is easily seen to be block diagonal with respect to β and α, all we need is the lower right p×p submatrix, which in view of the fundamental equality (2.42) can be written as

$$E\left[\frac{\partial L}{\partial \alpha} \cdot \frac{\partial L}{\partial \alpha'}\right].$$ (3.71)

There are various ways of consistently estimating (3.71), which will in general lead to different test statistics. Breusch and Pagan exploit the relationship

$$E\left(\frac{u_t^2}{\sigma^2} - 1\right)^2 = 2 \; , \tag{3.72}$$

which holds whenever the disturbances are normal. Since

$$\frac{\partial L}{\partial \alpha} = \frac{1}{2} \sigma^{-2} h'(\alpha_1) \Sigma z_t (u_t^2 \sigma^{-2} - 1) \tag{3.73}$$

under H_0, (3.71) can be written as

$$E\left(\frac{\partial L}{\partial \alpha} \frac{\partial L}{\partial \alpha'}\right) = \frac{1}{2} [\sigma^{-2} h'(\alpha_1)]^2 \Sigma z_t z_t' \; . \tag{3.74}$$

Evaluating this at $\hat{\alpha}_1$ and $\hat{\sigma}^2$ immediately yields the desired LM statistic

$$\xi = \frac{\partial L}{\partial \alpha'} (\hat{\beta}, \hat{\sigma}^2) \; \{E\left(\frac{\partial L}{\partial \alpha} \frac{\partial L}{\partial \alpha'}\right)\}^{-1} \frac{\partial L}{\partial \alpha} (\hat{\beta}, \hat{\sigma}^2)$$

$$= \frac{1}{2} (\Sigma z_t f_t)' (\Sigma z_t z_t')^{-1} (\Sigma z_t f_t) \; , \tag{3.75}$$

where $f_t = \hat{u}_t^2 / \hat{\sigma}^2 - 1$.

The statistic ξ can also be expressed in a different way by defining $g_t = f_t - 1 = \hat{u}_t^2 / \hat{\sigma}^2$, $g = (g_1, \ldots, g_T)'$, $f = (f_1, \ldots, f_T)'$, and $Z = (z_1, \ldots, z_T)'$. Then $f = g - i$, $i'g = T$, $i'f = f'Z(ZZ)^{-1}Z'i = 0$ and

$$\xi = \frac{1}{2} f'Z(Z'Z)^{-1}Z'f$$

$$= \frac{1}{2} (g'Z(Z'Z)^{-1}Z'g - (i'g)^2 / T) \; , \tag{3.76}$$

which is simply one half the explained sum of squares in the regression of g_t upon z_t. We therefore reject H_0 whenever the z_t's are able to "explain" much of the variation in the squared and standardized OLS residuals - a very suggestive decision rule indeed.

The test statistic ξ does not depend on the h-function that was used to specify the possible heteroskedasticity. The model (3.67) therefore comprises most types of heteroskedasticity as special cases, since h( ) is only restricted to possess first and second derivatives. Important special cases are for instance

$$\sigma_t^2 = \exp(z_t'\alpha) \; , \tag{3.77}$$

as considered by e.g. *Harvey* (1976), or

$$\sigma_t^2 = (z_t'\alpha)^m \tag{3.78}$$

with some prespecified integer m, as in *Goldfeld-Quandt* (1965) or *Glejser* (1969). The Breusch-Pagan result shows that LM-tests against such alternatives are asymptotically equivalent.

Another special case is the alternative considered by Harrison and McCabe, i.e. a discrete increase in σ_t^2 after T_1 observations. Here, Z is $T \times 2$, with $z_{2,t}$ equal to unity from 1 to T_1 and zero elsewhere, and h is the identity function. We have $\alpha = [\alpha_1, \alpha_2]'$, where $\alpha_2 = (\sigma_1^2 - \sigma_2^2)/\sigma_2^2 = -\theta$, the negative of the parameter introduced on page 34. From

(3.69), the derivative of the log-likelihood function with respect to α_2 can therefore be written as

$$\frac{\partial L}{\partial \alpha_2} = \frac{1}{2} \Sigma_{t=1}^{T} (\sigma^2 + \alpha_2) \left(\frac{u_t^2}{\sigma^4 (1 + \alpha_2)^2} - \frac{1}{(1 + \alpha_2) \sigma^2} \right) \tag{3.79}$$

which reduces to

$$\frac{\partial L}{\partial \alpha_2} (\alpha_2 = 0) = \frac{1}{2} \sigma^2 \Sigma_1^T \left(\frac{u_t^2}{\sigma^4} - \frac{1}{\sigma^2} \right)$$

$$= \frac{1}{2} \Sigma_1^T \left(\frac{u_t^2}{\sigma^2} - 1 \right) \tag{3.80}$$

under H_0.

Inserting the restricted ML estimates $\hat{u}_t$ and $\hat{\sigma}^2$ yields

$$\frac{\partial L}{\partial \alpha_2} (\alpha_2 = 0, \hat{u}) = \frac{1}{2} \Sigma_1^{T_1} \left(\frac{\hat{u}_t^2}{\hat{\sigma}^2} - 1 \right) = \frac{1}{2} \left(\frac{\hat{u}_1 {}' \hat{u}_1}{\hat{u} {}' \hat{u} / T} - T_1 \right)$$

$$= \frac{T_1}{2} \left(1 - \frac{\hat{u}_1 {}' \hat{u}_1 / T_1}{\hat{u} {}' \hat{u} / T} \right) , \tag{3.81}$$

where $\hat{u}_1$ corresponds to observations $1, \dots, T_1$. The expression (3.81) however is exactly the negative, as it should be, of the expression (3.66).

The statistic (3.75) can also be used to test for random variation in the regression coefficients. To see this, rewrite the equation as

$$y_t = \Sigma_{k=1}^{K} x_{tk} \beta_{tk} + u_t , \tag{3.82}$$

where $\beta_{tk} = \beta_k + v_{tk}$ and β_k is nonstochastic. Contrary to the standard model, v_{tk} is a random variable with expectation zero and variance σ_{tk}^2. The response of y_t to a unit change in x_{tk} therefore varies randomly around the mean response β_k. Such models are frequently entertained in the analysis of cross section data and were first considered by *Hildreth* and *Houck* (1968). Given the v_{tk}'s are both contemporaneously and serially uncorrelated with $E\, v_{tk} u_t = 0$, and given that $\sigma_{tk}^2 = \sigma_k^2$ does not vary over time, we can write (3.82) as

$$y_t = \Sigma_{k=1}^{K} x_{tk} \beta_k + \Sigma_{k=1}^{K} x_{tk} v_{tk} + u_t$$

$$= x_t {}' \beta + \epsilon_t , \tag{3.83}$$

where all assumptions of the standard linear regression model hold except that $\epsilon_t = \Sigma x_{tk} v_{tk} + u_t$ is now heteroskedastic, with variance

$$\sigma_t^2 = \sigma_u^2 + \Sigma_{k=1}^{K} x_{tk}^2 \sigma_k^2 . \tag{3.84}$$

This type of heteroskedasticity is of the form (3.67) with $z_{tk} = x_{tk}^2$ and h the identity function.

Testing for arbitrary heteroskedasticity

A related test, where similar to the Breusch-Pagan approach the type of the possible heteroskedasticity need not be specified a priori is due to *White* (1980). In particular, the White test also covers the Hildreth-Houck random coefficients alternative. The basic idea is to compare the standard OLS based estimator of the covariance matrix of the OLS co-efficient estimates $\hat{\beta}$,

$$\hat{V}(\hat{\beta}) = \hat{\sigma}^2 (X'X)^{-1} , \qquad (3.85)$$

to an alternative which is consistent under heteroskedasticity. One such alternative is shown by White to be

$$\tilde{V}(\hat{\beta}) = (X'X)^{-1} (\Sigma_{t=1}^{T} \hat{u}_t^2 x_t x_t')(X'X)^{-1} . \qquad (3.86)$$

Given homoskedastic disturbances and assumption (2.3),

$$\lim_{T \to \infty} T\hat{V}(\hat{\beta}) = \lim_{T \to \infty} T\tilde{V}(\hat{\beta}) = \sigma^2 Q^{-1} , \qquad (3.87)$$

which means that the expressions (3.85) and (3.86) should be close to each other without heteroskedasticity. Under heteroskedasticity, on the other hand, $T\hat{V}(\hat{\beta})$ and $T\tilde{V}(\hat{\beta})$ will in general tend to different probability limits, so an obvious test can be based on the difference.

The resulting test statistic is rather cumbersome to compute. However, a comparable statistic with the same limiting distribution is easily obtained from the artificial regression

$$\hat{u}_t^2 = \alpha_0 + \Sigma_{i=1}^{K} \Sigma_{j=i}^{K} \alpha_{ij} x_{ti} x_{tj} , \qquad (3.88)$$

that is, from a regression of the squared OLS residuals on all second order products and cross products of the initial regressors. White shows that T times R^2 (constant adjusted) from the regression (3.88) is asymptotically equivalent to his initial test statistic.

The statistic TR^2 is asymptotically χ^2 under H_0, with degrees of freedom equal to the number of linear independent regressors in (3.88). There will be redundancies if the initial regression contains a constant and polynomial terms. For example if $x_{t1} = 1$ and $x_{t3} = x_{t2}^2$, then $x_{t1} x_{t3} = x_{t2} x_{t2}$. The degrees of freedom are $K(K+1)/2$ when there are no redundancies.

The test is consistent against any alternative which leads to different probability limits for $T\hat{V}(\hat{\beta})$ and $T\tilde{V}(\hat{\beta})$, respectively. This can happen also when the true source of the trouble is something other than heteroskedasticity. Contrary to e.g. *Hall* (1983), White's test can therefore also to be used to test for misspecification of the deterministic part of the model.

For illustration, consider the case where $K = 1$, $y_t = x_t \beta + u_t$ and

$$x_t = [1 + (-1)^t]/2$$

(i.e. the independent variable is a sequence of ones and zeros). Assume that the true relationship contains a constant c, but that the regression is estimated without the intercept. Then it is straightforwardly verified that

$$\text{plim}_{T \to \infty} \; \hat{\sigma}^2 = \sigma^2 + c^2/2 \; , \tag{3.89}$$

so

$$\text{plim}_{T \to \infty} \; T\hat{V}(\beta) = \text{plim}_{T \to \infty} \; T\hat{\sigma}^2 (X'X)^{-1} = 2\sigma^2 + c^2 \tag{3.90}$$

On the other hand,

$$\text{plim}_{T \to \infty} \; T\tilde{V}(\beta) = 2 \, \sigma^2 \; , \tag{3.91}$$

which differs from plim $T\hat{V}(\beta)$ by c^2. The White test will therefore also detect the omitted constant.

The White test is also closely linked to the Breusch-Pagan test. To see this, simply use products and cross products of regressors as the auxiliary z-variables in (3.67). The Breusch-Pagan test statistic is then

$$\xi = f'Z(Z'Z)^{-1}Z'f/2$$
$$= W'Z(Z'Z)^{-1}Z'W/2\hat{\sigma}^4 \; , \tag{3.92}$$

where $w_t = f_t \hat{\sigma}^2 = \hat{u}_t^2 - \hat{\sigma}^2$, as compared to White's test statistic

$$\xi^* = W'Z(Z'Z)^{-1}Z'W/\frac{1}{T}W'W \; . \tag{3.93}$$

Under H_0, and given that the disturbances are normal, we have

$$E(u_t^4) = 3\sigma^4 \; , \tag{3.94}$$

or equivalently,

$$\text{var}(u_t^2) = 2\sigma^4 \; , \tag{3.95}$$

which implies

$$\text{plim}_{T \to \infty} \; 2\hat{\sigma}^4 = \text{plim}_{T \to \infty} \; \frac{1}{T}W'W = 2\sigma^4 \; . \tag{3.96}$$

The Breusch-Pagan and White tests are therefore asymptotically equivalent, since (3.96) can be shown to hold for arbitrary local alternatives. The White test however retains its prescribed null distribution also under nonnormal disturbances, provided only that $E(u_t^4) < \infty$, whereas the Breusch-Pagan test statistic ξ will then no longer have an asymptotic χ^2 distribution. This follows from the relationship (3.72), which holds only for the normal distribution and distributions with kurtosis like the normal. We thus have one of the rare instances where even the asymptotic distribution of a test statistic depends on the normality of the disturbances. This has lead *Koenker* (1981) to "studentize" the Breusch-Pagan test by substituting $W'W/2T$ for $\hat{\sigma}^4$ in (3.92). This modified test will then have an asymptotic χ^2 distribution under H_0, whether the disturbances are normal or not.

Since both the Breusch-Pagan and the White tests do not require specifying the heteroskedastic structure of the disturbances, they might have less power than other tests for any given particular alternative. On the other hand, it is difficult enough to specify the regression part of the model, about which there is generally some guidance, without being required to state exactly how the disturbance variances change, which makes these tests attractive for applications where no particular form for the heteroskedasticity suggests itself.

c) Normality

Most of the tests considered in this monograph require normality of the regression disturbances as part of the respective null hypothesis. There are two reasons for this. First, the finite sample null distribution of many tests, as well as their optimality in some sense, can only be derived when the disturbances are normal. Otherwise, the tests will in general attain their nominal size only asymptotically. Second, normality of the u_t's is a basic prerequisite for all tests that are based on either the LM, LR or Wald principles. Very often, such tests attain their nominal significance level only asymptotically even under normality of the u_t's, due to unmanageable finite sample distributions of the respective test statistics (a case in point being the Breusch-Pagan and White tests for heteroskedasticity discussed above). Departure from normality will then affect only the asymptotic optimality properties of the tests and does not introduce any additional finite sample problems.

Very rarely, as for the unmodified Breusch-Pagan test, do nonnormal disturbances affect the asymptotic null distribution of a test statistic.

Another reason why one might wish to know whether or not the regression disturbances are normal concerns the efficiency of the OLS parameter estimates $\hat{\beta}$. These are best unbiased under normality, and best only in the much smaller class of *linear* unbiased estimators otherwise. This implies that there are nonlinear estimators which outperform OLS when the disturbances are nonnormal. For instance, the ratio of the variance of the ML estimator to $\text{var}(\hat{\beta}_i)$ drops to zero as $T \to \infty$ when the disturbances follow a rectangular distribution (see e.g. *Harvey*, 1981, pp. 113-115). At the other extreme of heavy-tailed distributions, OLS is outperformed by Least-Absolute-Deviations or similar robust procedures, so interest in the disturbance distribution is justified also from this efficiency viewpoint. We leave it however to the reader to determine the seriousness of this problem as compared to what else could happen to his model.

Comparing moments

There are various characteristics of the normal distribution which could form the basis of a test. Many procedures compare sample order statistics   to what is expected under normality. Best known among these is the Shapiro-Wilk statistic

$$W = \frac{\sum_{i=1}^{h} a_{iT} (u_{(T-i+1)} - u_{(i)})^2}{\sum_{i=1}^{T} u_i^2} \tag{3.97}$$

where $u_{(1)} \le u_{(2)} \le ... \le u_{(T)}$ are the ordered regression disturbances, and where $h = T/2$ or

$h = (T-1)/2$ according to whether T is even or odd. The a_{iT} are coefficients tabulated by *Shapiro* and *Wilk* (1965). The critical region is given by low values of W, and appropriate significance points are also provided by Shapiro and Wilk (1965).

Below we focus on methods based on the moments of the disturbances. First we proceed as if the u_t's were observed.

Since under normality (and more general for any symmetric distribution whose respective moments exist)

$$E(u_t^n) = 0 \text{ (n odd)}, \tag{3.98}$$

an obvious approach is to determine whether the corresponding sample moments deviate significantly from zero. In particular, significance points for the statistic

$$\sqrt{b_1} = \frac{1}{T} \Sigma_{t=1}^{T} (u_t - \bar{u})^3 / S^3 , \tag{3.99}$$

where S is the square root of $S^2 = \Sigma(u_t - u)^2/T$, have been tabulated by *Pearson* and *Hartley* (1966). (Most readers will recognize that (3.99) is the sample analogue of the skewness $\gamma = E(u_t^3)/\sigma^3$ of the disturbance distribution).

One can also exploit a well known fact about the kurtosis of the u_t's, i.e. that under normality,

$$E(u_t^4)/\sigma^4 = 3 . \tag{3.100}$$

(see also (3.94) and (3.95)).

Again, significance points of the sample analogue

$$b_2 = \frac{1}{T} \Sigma_{t=1}^{T} (u_t - \bar{u})^4 / S^4 \tag{3.101}$$

can be found in *Pearson* and *Hartley* (1966).

Since each of the b_1 and b_2 tests will detect only one directional departures from normality, one might also consider the joint use of $\sqrt{b_1}$ and b_2. If $\sqrt{b_1}$ and b_2 were independent, a joint test at level α would reject H_0 whenever $(\sqrt{b_1}, b_2)$ fell outside a rectangle determined by the two individual α' rejection regions, where

$$\alpha' = (1-(1-\alpha)^{1/2})/2 . \tag{3.102}$$

Thus, if either $\sqrt{b_1}$ or b_2 lies beyond its upper or lower $100\alpha'$ points, this joint test rejects the null hypothesis of normality at the $\alpha\%$ significance level. *Pearson, D'Agostino* and *Bowman* (1977) have elaborated this idea by adjusting α' so as to compensate for the lack of complete independence of $\sqrt{b_1}$ and b_2.

Following *Jarque* and *Bera* (1980), one can alternatively combine b_1 and b_2 into the statistic

$$JB = T[\frac{1}{6} b_1 + \frac{1}{24} (b_2 - 3)^2] \tag{3.103}$$

This particular combination results from viewing the normal distribution as a member of the Pearson Family of distributions (see e.g. *Kendall* and *Stuart*, 1969, p. 148) and from applying the LM principle to test whether the parameters which determine the normal distribution have the required size. The normality hypothesis is rejected whenever JB is too large, with the appropriate rejection region given by the asymptotic χ^2 null distribution of JB.

Residuals vs. true disturbances

So far, we have assumed that the the the true disturbances u_t are known. What if the u_t's are replaced by the OLS-residuals $\hat{u}_t$?

Since the $\hat{u}_t$'s are under H_0 neither independent nor homoskedastic, the null distributions of the various test statistics will certainly be affected. However, *White* and *MacDonald* (1980) show that this does not matter asymptotically, i.e. all tests have correct size α as $T \to \infty$.

Residual related problems with the null distribution can be avoided by using recursive or BLUS residuals, which are NID$(0,\sigma^2)$ under H_0. The null distribution of the test statistics is then exactly as if T-K disturbances were actually observed. On the other hand, the power of the tests will then be considerably below what is obtained with OLS residuals (*Huang* and *Bolch*, 1974; *Ramsey*, 1974).

This loss of power is the second issue involved in replacing the true disturbances by estimated residuals of whatever kind. Though the loss appears to be smallest for the OLS residual vector $\hat{u}$, it can still be considerable even then. *Weisberg* (1980) has pointed out that the $\hat{u}_t$'s, being linear combinations of the u_t's, have for this reason a tendency to appear normal even when the u_t's are not. The coefficients of these linear combinations are given by the rows of the matrix $M = (m_{ij}) = I - X(X'X)^{-1}X'$ from (2.8), i.e.

$$\hat{u}_t = \Sigma_{i=1}^T m_{ti} u_i .$$

(3.104)

Assumption (2.3) guarantees that

$$m_{tt} \to 1, \quad \Sigma_{i \neq t} m_{ti}^2 \to 0 \quad (t = 1,...,T)$$

(3.105)

as $T \to \infty$, so the relationship (3.104) need not bother us asymptotically. *Weisberg* (1980) however gives finite X matrices where the coefficients in (3.104) are of roughly the same size for all t and i and where the power of most tests is very poor.

We do not enter into further detail here and refer the reader to *Mardia* (1980) for a comprehensive survey of normality tests.

4. TESTING REGRESSORS

We now turn to procedures for detecting irregularities in the deterministic part of a linear regression model. The various alternatives one might consider here are to some extent overlapping (in the sense that a non-zero mean of the regression disturbances can either be viewed as due to an omitted regressor, or as the consequence of a wrong functional form, or as some type of instability in the regression coefficients), so these techniques do no more group unambiguously into classes according to the respective alternative.

In fact, we will see in chapter 5 below that almost all procedures can be viewed as tests for omitted variables. The classification we adopt here is therefore rather arbitrary.

a) Structural Change

The problem of structural shift in regression arises in various fields. In the study of growth in biology, for instance, one often assumes a log-linear relationship between the size of two body parts, and that this relationship persists throughout stable growth periods. Structural shifts here indicate that a new phase has begun and may therefore be of considerable interest. In economics, there is often reason to suspect that some model parameters have changed due to political events (change of government, new tax laws etc.). One cannot for instance be confident that key parameters like the marginal propensity to consume were the same in the Great Depression as in the 1960's, as is often done in empirical studies.

The common feature of such situations is a natural ordering of the data points, either by size, age or, as we are assuming here, by time. The respective model is assumed to fit the first segment of the data, and the problem is to test whether this segment covers the whole data set.

The Chow test and related methods

We consider first the simplest case where there is just one possible shift in the regression parameters, and where in addition the shift point T^* is known. Moreover, we assume initially that $T^* \geq K$, $T-T^* \geq K$, $T > 2K$, and that the regressor submatrices $X_1 = [x_1, ..., x_{T^*}]'$ and $X_2 = [x_{T^*+1}, ..., x_T]'$ have both full column rank. The regression model (2.2) can then in obvious notation be split into

$$y_1 = X_1\beta_1 + u_1 \text{ and } y_2 = X_2\beta_2 + u_2 , \tag{4.1}$$

where the null hypothesis of parameter constancy is equivalent to $\beta_1 = \beta_2 = \beta$.

The following test is due to *Chow* (1960): Fit both parts of (4.1) separately and reject H_0 whenever

$$F = \frac{(\hat{u}'\hat{u} - \tilde{u}'\tilde{u})/K}{(\tilde{u}'\tilde{u})/(T-2K)} \tag{4.2}$$

is too large, where $\tilde{u} = [\hat{u}_1', \hat{u}_2']'$ is the vector of OLS residuals from fitting both parts of (4.1) separately, and $\hat{u}$ is the standard OLS residual vector from the overall regression. This is intuitively very reasonable, since large values of F imply that the OLS fit is greatly increased by breaking up the regression. The exact rejection region is determined by noting that under H_0, F has an F distribution with K and T-2K degrees of freedom.

The Chow test is closely related to the Likelihood Ratio (LR) test of the hypothesis $H_0: \beta_1 = \beta_2$. This can be seen as follows:

Under H_0, the log likelihood function of the sample (4.1) is (compare 2.27)

$$L(\beta, \sigma^2, y) = -\frac{T}{2} \ln(2\pi) - \frac{T}{2} \ln \sigma^2 - \frac{(y_i - X_i\beta_i)'(y_i - X_i\beta_i)}{2\sigma_i^2}$$

The Maximum Likelihood estimates for β and σ^2 are $\hat{\beta} = (X'X)^{-1}X'y$ and $\hat{\sigma}^2 = \hat{u}'\hat{u}/T$, respectively. The maximized value of the log likelihood function is therefore

$$l_0 = -\frac{T}{2} \ln(2\pi) - \frac{T}{2} - \frac{T}{2} \ln(\hat{\sigma}^2) .$$

Under H_1 (that is, when β_1 is possibly different from β_2, but the disturbance variances are still identical across subsamples), the ML estimates for β_1 and β_2 are obtained by fitting both parts of (4.1) separately, with residual vectors $\hat{u}_1$ and $\hat{u}_2$, and the ML estimate for σ^2 is $\tilde{\sigma}^2 = \tilde{u}'\tilde{u}/T$. The maximised value of the log likelihood function is therefore

$$l_1 = -\frac{T}{2} \ln(2\pi) - \frac{T}{2} - \frac{T}{2} \ln(\tilde{\sigma}^2) .$$

The test statistic for the LR test is then

$$LR = T [\ln(\hat{\sigma}^2) - \ln(\tilde{\sigma}^2)] ,$$

a monotone transformation of (4.2). LR is asymptotically χ_K^2 under H_0, and would produce the same rejection region as (4.2) if its exact finite sample null distribution were known.

When $T^* > T\text{-}K$ (i.e. when the parameter shift occurs very late in the sample), the second regression in (4.1) cannot be computed due to rank deficiency in X_2. For such cases, Chow suggests rejecting H_0 for large values of

$$C = \frac{(\hat{u}\,'\hat{u}\text{-}\hat{u}_1\,'\hat{u}_1)/(T\text{-}T^*)}{(\hat{u}_1\,'\hat{u}_1)/(T^*\text{-}K)} \quad . \tag{4.3}$$

Under H_0, C has again an F distribution, with $T\text{-}T^*$ and $T^*\text{-}K$ degrees of freedom. C can of course also be computed when $T^* < T\text{-}K$, but the resulting test is then less powerful than the test based on (4.2). When $T^* = T\text{-}K$, $\hat{u}_2 = 0$, i.e. $\tilde{u}\,'\tilde{u} = \hat{u}_1\,'\hat{u}_1$ and C and F are identical. When $T^* < K$ (i.e. when the parameter shift occurs very early in the sample), the analogue of (4.3) applies with $\hat{u}_2$ in place of $\hat{u}_1$. The Chow test cannot be applied when both $T^* \leq K$ and $T\text{-}T^* \leq K$.

The form (4.3) of the Chow test can be viewed as a test for predictive failure. To this purpose, consider the model

$$\begin{bmatrix} y_1 \\ y_2 \end{bmatrix} = \begin{bmatrix} X_1 & 0 \\ X_2 & I \end{bmatrix} \begin{bmatrix} \beta_1 \\ \delta \end{bmatrix} + \begin{bmatrix} u_1 \\ u_2 \end{bmatrix} \tag{4.4}$$

where $\delta = Ey_2\text{-}X_2\beta_1$ is a $(T\text{-}T^*)$ vector of unknown coefficients, which equal the mean errors when predicting y_2 from the first period. The hypothesis that the mean prediction errors are all zero therefore amounts to $H_0^* : \delta = 0$.

In the model (4.4), a dummy variable is included for every observation in the second period. All the second period residuals will therefore be zero, and the OLS-estimate for β in the model (4.4) will be exactly the same as is obtained from running the first period regression alone. This implies that the standard F-test for H_0^* has the test statistic (4.3) (compare 2.35). In addition, the OLS estimates $\hat{\delta}$ from (4.4) are the vector of empirical prediction errors. The individual t-statistics can therefore be used to test period by period whether each prediction error differs significantly from zero.

Testing subsets of the regression coefficients

The model (4.1) can be reparameterized as

$$\begin{bmatrix} y_1 \\ y_2 \end{bmatrix} = \begin{bmatrix} X_1 & 0 \\ X_2 & Z \end{bmatrix} \begin{bmatrix} \beta_1 \\ \beta_2\text{-}\beta_1 \end{bmatrix} + \begin{bmatrix} u_1 \\ u_2 \end{bmatrix} \tag{4.5}$$

where $Z = X_2$. The null hypothesis of parameter constancy is here equivalent to H_0: $\beta_2\text{-}\beta_1 = 0$, and can be tested with the standard F-test. From the general form (2.35) of the F-test it is clear that this produces the test statistic (4.2).

Due to the form of (4.5), the Chow test has in the statistics literature for some time been known as the Analysis of Covariance test.

The reparameterized model (4.5) also shows how to generalize the Chow test to situations where only part of the regression coefficients are subject to a possible shift. This is done by including in Z only those regressors that correspond to the parameters under test. Similarly, extensions to two or more shifts are straightforward, given that all switchpoints are known, by writing the regression as e.g.

$$
\begin{bmatrix} y_1 \\ y_2 \\ y_3 \end{bmatrix} = \begin{bmatrix} X_1 & 0 & 0 \\ X_2 & X_2 & 0 \\ X_3 & 0 & X_3 \end{bmatrix} \begin{bmatrix} \beta_1 \\ \beta_2 - \beta_1 \\ \beta_3 - \beta_1 \end{bmatrix} = \begin{bmatrix} u_1 \\ u_2 \\ u_3 \end{bmatrix}
$$

and testing for $\beta_2 - \beta_1 = \beta_3 - \beta_1 = 0$.

The prior knowledge that only a subset of the regression parameters is subject to a possible shift of course improves the power of the test. A related type of prior knowledge arises when all regressors are monotone functions of time, as often happens in economic applications, and when in addition the expected value of y_t is a continuous function of x_t even under the alternative. This imposes the restriction

$$
x'_T \cdot (\beta_2 - \beta_1) = 0 , \tag{4.6}
$$

which can then be used to improve estimation under the alternative. This situation is for the bivariate regression model illustrated in Figure 4.1, where in the lower panel Ey_t is a continuous function of x even under the alternative, while in the upper panel it is not.

Disregarding the restriction (4.6) and applying the standard Chow test is of course still a valid, though less powerful procedure.

Coefficient stability vs.stability of the variance

There exists considerable confusion among econometricians concerning both the null and alternative hypotheses of the Chow test.

We have shown above that the Chow test is equivalent to the LR test *when the disturbance variance remains constant even under the alternative.* (In fact, the Chow test can then be shown to be uniformly most powerful invariant (UMPI)). This requirement will however almost never be met in practice, since a shift in the regression coefficients is likely to be accompanied by a shift in the variance as well. Whenever this happens, the Chow test is no more the LR test, and there will in general exist more powerful procedures. Of course, the Chow test is still a valid test.

If both coefficients and variances differ across the subsamples in (4.1), the maximized value of the log likelihood function is

$$
l_1^* = -\frac{T}{2}\ln(2\pi) - \frac{T}{2} - \frac{T^*}{2}\ln\hat{\sigma}_1^2 - \frac{(T-T^*)}{2}\ln\hat{\sigma}_2^2 ,
$$

FIG. 4.1: CONTINUITY VS. DISCONTINUITY AT THE SHIFT POINT

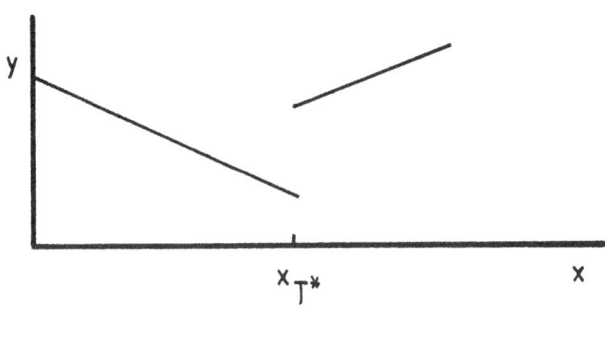

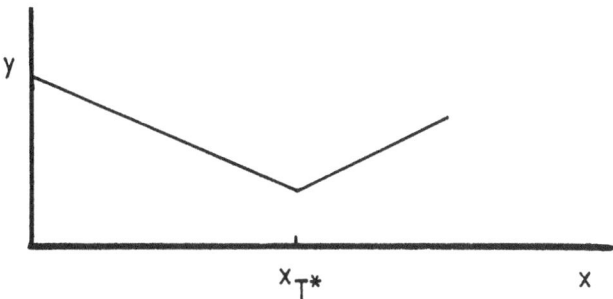

where $\hat{\sigma}_1^2 = \hat{u}_1{}'\hat{u}_1/T^*$ and $\hat{\sigma}_2^2 = \hat{u}_2{}'\hat{u}_2/(T\text{-}T^*)$. This produces the LR test statistic

$$LR^* = 2(l_0\text{-}l_1^*),$$

which is asymptotically χ^2_{K+1} under H_0 and   which   implies a different rejection region than the Chow test. In particular, LR leads to a more powerful test over some regions of the parameter space. No test is however uniformly superior to the other.

A different issue often confused with this first one concerns a reformulation of the null hypothesis. Often, one is mainly interested in the regression coefficients, and wants to determine whether they are constant, regardless of what happens to the variance. The hypothesis under test is thus H_0^*: $\beta_1 = \beta_2$, where σ_1^2 and σ_2^2 are allowed to *differ under the null hypothesis* (Statisticians will note that this is the regression analogue of the well known Behrens-Fisher problem of testing the difference between two means).

Obviously, if the null hypothesis is put like this, the Chow test is no longer valid. *Jayatissa* (1977) has suggested an alternative which is valid even for the enlarged null hypothesis, but hard to compute and not uniformly most powerful. Related extensions are due to *Watt* (1979) and *Honda* (1982). *Anderson* and *Mizon* (1983) and *Pesaran et al.* (1985) provide general discussions of the statistical problems surrounding the various specifications of the null and alternative hypotheses. The latter paper in addition extends the standard tests to dynamic models and simultaneous equation systems. A good introduction to the general Behrens-Fisher problem can also be found in *Kendall* and *Stuart* (1967). *Quandt* (1958, 1960) has first introduced the statistic LR^* in econometrics. For this reason, the expressions

$$
\lambda_{T^*} = \ln \left(\frac{\text{max likelihood given } H_0}{\text{max likelihood given } H_1} \right)
$$

$$
= \frac{T^*}{2} \ln \hat{\sigma}_1^2 + \frac{(T-T^*)}{2} \ln \hat{\sigma}_2^2 - \frac{T}{2} \ln \hat{\sigma}^2 \tag{4.7}
$$

are often called Quandt Ratios, and the resulting test is referred to as the Quandt test.

Salkever (1976) and *Dufour* (1982) give a more detailed account of the Chow test's role as a prediction error test. These results are generalized to dynamic and simultaneous models by *Pagan* and *Nicholls* (1984).

Shift point unknown

A major drawback of both the Chow and Quandt tests is the requirement that the shift point T^* be known. If we keep the assumption that there is at most one shift point, at an unknown point in time, one could still proceed with the standard Chow test by setting $T^* = T/2$ (for T even; otherwise take the nearest integer). The true shift point, if any, would then either be to the left or to the right of $T/2$, or be equal to $T/2$. In the latter case, there is no problem at all. If $T/2$ differs from the true T^*, the fit of the regression can still be improved by breaking it into parts, although to a lesser extend.

Farley and *Hinich* (1970) suggest an alternative based on (4.5). When T^* is known, the Chow test   checks whether $\delta = \beta_2 - \beta_1 = 0$, where the   t'th row of the regressor matrix corresponding to δ is given by

$$
z_t(T^*) = 0 \quad (1 \leq t \leq T^*) \text{ and}
$$

$$
z_t(T^*) = x_t \quad (T^* + 1 \leq t \leq T) . \tag{4.8}
$$

When T^* is unknown,   they   suggest to approximate $z_t(T^*)$ by its average value over all T^*'s $(T^* = 1,...,T)$, i.e. $z_t = tx_t$, and then again to test whether $\delta = 0$. This procedure and the pseudo Chow test based on setting $T^* = T/2$ have identical distributions under H_0, but the Farley-Hinich test has more power, ceteris paribus, for large values of $[T^* - T/2]$ (see *Farley* et al., 1975).

Still another approach is to let the data determine the most likely shift point (which when σ^2 is allowed to differ across regimes and $K < T^* < T-K$ is given by the values of T^* where the Quandt-Ratio λ_{T^*} from (4.7) attains its minimum) and then to perform a Likelihood Ratio test. Note however that the likelihood function here is not differentiable with respect to the parameter T^*, a problem that does not arise when T^* is known. In particular, the asymptotic distribution theory one usually invokes to demonstrate that -2 log λ_{T^*} has an asymptotic χ^2 distribution under H_0 does, for this reason, not apply here. In fact, *Quandt* (1960) has found that the null distribution of -2 log λ_{T^*} is markedly non-χ^2 whenever T^* is estimated from the data. The likelihood ratio approach to the present problem therefore suffers from a lack of knowledge of the appropriate rejection region, and only recently has this deficiency been partly overcome for very special X-matrices (*Hawkins*, 1977; *Worsley*, 1979).

The problem of testing for structural shifts is closely related to what in econometrics is known as "switching regression" (see e.g. *Goldfeld* and *Quandt*, 1973, or *Poirier*, 1973). Our discussion above was concerned with the case where the switch from one regime to another occurs on the basis of the time index. This however is only one among a variety of possible indicators for structural change. Others could be whether or not a certain variable exceeds some threshold limit, or whether, when studying the quantity of some commodity traded on a market in disequilibrium, we observe a demand or a supply function (*Fair* and *Jaffee*, 1972). We will not go into detail here. A good reference for further study is *Poirier* (1976). An extensive bibliography of the literature on the change-point problem in general has also been compiled by *Hinkley* et al. (1980) or *Hackl* and *Westlund* (1985).

The CUSUM and CUSUM of squares tests

We now consider situations where neither the number nor the timing of possible shifts in the regression parameters are known. A well known procedure to detect such types of deviations from the standard linear regression model, which has since gained wide acceptance in empirical econometrics and elsewhere, is the CUSUM test by *Brown* et al. (1975).

The CUSUM test is based on recursive residuals, as defined on p. 7 above. Remember that each recursive residual $\tilde{u}_t$ represents the (standardized) discrepancy between the actual value of the dependent variable at time t and the optimal forecast using only the information contained in the previous t-1 observations. The basic idea is that if β is constant up to time $t = r$ and differs from that constant from then on, the recursive residuals $\tilde{u}_t$ will have zero mean for t up to r, but will in general have nonzero means subsequently. Thus it seems reasonable to expect some information on possible structural changes from a plot of the CUSUM quantity

$$W_r = \frac{1}{s} \Sigma_{t=K+1}^r \tilde{u}_t \qquad (4.9)$$

against r for $r = K+1, \ldots, T$, where the scale factor $s = (\hat{u}'\hat{u}/(T-K))^{1/2}$ is an estimate for the standard deviation of the disturbances.

Here the great advantage of recursive residuals becomes apparant, as compared to OLS and BLUS residuals: A parameter shift at $t = r$ will in general induce nonzero means for OLS and BLUS residuals also prior to r while recursive residuals prior to r will not be affected at all. Thus recursive residuals are particularly convenient whenever it comes to testing the constancy of regression relationships over time .

Let us now look more closely at the recursive residuals when parameters are unstable, i.e. when the basic regression relationship (2.1) is more generally written as

$$y_t = x_t'\beta_t + u_t ,$$ (4.10)

where the β_t's may now differ across observations.

Let C be the $(T-K) \times T$ matrix, with t'th row as in (2.18), that is implicitly used when computing the vector $\tilde{u} = Cy = Cu$ of recursive residuals. Note first that the normality of $\tilde{u}$ is guaranteed even if H_0 does not hold (given that instability of the regression parameters is the only deviation from the classical assumptions), since $\tilde{u}$ is a linear transformation of u. Second, if the variance of the u_t's remains constant from $t = 1, \ldots, T$, we have from

$$\text{cov}(\tilde{u}) = E(Cuu'C') = \sigma^2 CC' = \sigma^2 I_{T-K}$$ (4.11)

that independence of the recursive residuals is also preserved, whether the β_t's are stable or not (if the disturbance variance varies, however, we will have dependence as well as heteroskedasticity among the recursive residuals). Any non-constancy of the β_t's will thus only affect the central tendency of the $\tilde{u}_t$'s.

The expected values of $\tilde{u}_t$ $(t = K+1, \ldots, T)$ may follow a variety of patterns depending on the trajectories of β_t and x_t. Using the notation of section 2.a, we have from (2.16) and (2.17) that

$$\tilde{u}_t = (y_t - x_t'\hat{\beta}^{(t-1)})/f_t ,$$ (4.12)

where $\hat{\beta}^{(t-1)}$ is the OLS estimate for β based on the first t-1 observations. Thus

$$E(\tilde{u}_t) = (x_t'\beta_t - x_t'E\hat{\beta}^{(t-1)})/f_t$$
$$= x_t'[\beta_t - (X^{(t-1)'}X^{(t-1)})^{-1} \sum_{i=1}^{t-1} x_i x_i'\beta_i]/f_t .$$ (4.13)

If there is just one regressor (K = 1), this takes the form

$$E(\tilde{u}_t) = x_t \sum_{i=1}^{t} a_i^{(t)}\beta_i/f_t ,$$ (4.14)

where $a_t^{(t)} = 1$ and $a_i^{(t)} = -x_i^2/\sum_{i=1}^{t-1} x_i^2$ $(i = 1, \ldots, t-1)$, i.e. $\sum_{i=1}^{t} a_i^{(t)} = 0$.

From this it is immediately apparent that $E(\tilde{u}_t) = 0$ whenever $\beta_1 = \ldots = \beta_T = \beta$, whereas in general $E(\tilde{u}_t) \neq 0$ otherwise. If for instance β_t suddenly jumps at time r, and again remains constant from then on, one has for $t > r$

$$E(\tilde{u}_t) = x_t \Delta\beta \sum_{i=r+1}^{t} a_i^{(t)}/f_t \neq 0$$ (4.15)

Since $\Sigma_{i=r+1}^{t} a_i^{(t)} > 0$ for all r, we have thus in particular that $E(\bar{u}_t) > 0$ when both the x_t's and $\Delta\beta = \beta_{r+1} - \beta_r$ are positive. The same happens if $\Delta\beta < 0$ and $x_t < 0$ ($t = r+1,...,T$). When the x_t's and $\Delta\beta$ have different signs, the expectation $E(\bar{u}_t)$ will be negative.

For multiple regressions ($K > 1$) the situation is more complex. The expression in parantheses in (4.13) is then a matrix-weighted average of the β_t's ($t = 1,...,r$), and $E(\bar{u}_t)$ then in addition depends on the angle between $\Delta\beta$ and this average. In particular, if these are orthogonal, $E(\bar{u}_t) = 0$ irrespective of any parameter instability. One can thus easily construct trajectories of β_t and x_t ($t = 1,...,T$) such that $E(\bar{u}_t) = 0$ for all $t = K+1,...,T$, in which case the CUSUM test will have no power (a point to be reconsidered in our Monte Carlo experiments below).

The CUSUM quantity (4.9) has expectation $E(W_r) = 0$ ($r = K+1,...,T$) under H_0. This follows immediately from the fact that W_r is an odd function of the true disturbance vector u. If however the interplay of parameter variation and regressors is such that the recursive residuals show a tendency to be positive (or negative) over sufficiently long sub-periods, or if few big prediction errors occur, W_r may be expected to be large in absolute value.

Whether or not a departure of W_r from the mean value line $E(W_r) = 0$ is significant can be decided in a number of ways. Brown et al. propose a pair of lines lying symmetrically above and below the line $W_r = 0$, such that under H_0 the probability of crossing one or both lines equals the required significance level α. This is graphed in Figure (4.2), together with a typical time path of the CUSUM quantities W_r. Here, W_r crosses the critical line at T^*, indicating some parameter instability prior to T^*. In fact, a good guess for the switch point is where the CUSUM plot starts wandering off in either direction.

FIG. 4.2: SAMPLE PATH OF CUSUM STATISTIC

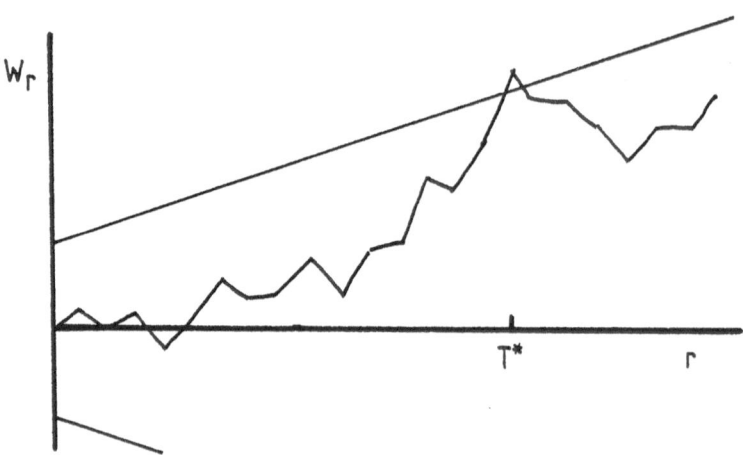

In order to determine the appropriate critical lines, given some significance level α, let us instead of (4.9) consider

$$\tilde{W}_r = \frac{1}{\sigma} \Sigma^r_{t=K+1}\, \tilde{u}_t = \frac{s}{\sigma}\, W_r \,. \tag{4.16}$$

The $\tilde{W}_r$'s differ from the W_r's in that s is replaced by its population analogue σ. This does not matter asymptotically, since it is immediate from (4.16) that $\text{plim}_{T\to\infty}\, \tilde{W}_r\text{-}W_r = 0$ for all r, but greatly facilitates the small sample analysis. It is easily seen that the W_r's are jointly normally distributed, with

$$E(\tilde{W}_r) = 0,\ E(\tilde{W}_r^2) = r - K,\ E(\tilde{W}_r\tilde{W}_s) = \min(r,s) - K \,. \tag{4.17}$$

Following Brown et al., we further simplify the analysis by assuming that the $\tilde{W}_r$'s are discrete observations of a continuous Gaussian process $Z_t(K \leq t \leq T)$ with mean and covariance functions as in (4.17). This is in fact the Brownian motion process starting from zero at time $t = K$. The simplification arises because this process, and in **particular** its crossing probabilities, have been studied extensively in probability theory, providing us with all the results we need.

Note however that if one wished to find a curve such that under H_0 the probability that the sample path lies above the curve is the same for all r ($K \leq r \leq T$), one should choose curves of the form

$$f(r) = \pm\lambda(r\text{-}K)^{1/2} , \tag{4.18}$$

where λ is some positive constant. This is immediate from (4.17), i.e. the fact that the standard deviation of the $\tilde{W}_r$'s is $(r\text{-}K)^{1/2}$. Consequently, if the analysis is limited to straight lines, the crossing probability cannot be constant for all r. On the other hand, the probability that the process crosses the curve at least once is much easier to determine for straight lines, using the Brownian motion approximation.

Even after confining oneself to straight lines, however there is no obvious rule how to proceed. The optimal choice will in general depend on the alternative and any information one has as to the structure and timing of possible parameter shifts.

In the absence of such prior knowledge, Brown et al. propose choosing lines such that under H_0, the crossing probability is maximized for r halfway between T and K. This implies that the line should be tangent to the curve (4.18) at $r = (T + K)/2$, i.e. have slope

$$f'((T+K)/2) = \pm\frac{\lambda}{2}((T+K)/2)^{-1/2} . \tag{4.19}$$

Given λ, the line is thus uniquely determined by (4.19) and the requirement that it pass through $\{(T+K)/2, f((T+K)/2)\}$. Straightforward computation shows that this leads to a pair of lines through the points

$$\{K, \pm a(T\text{-}K)^{1/2}\}, \{T, \pm 3a(T\text{-}K)^{1/2}\} , \tag{4.20}$$

where $a = \lambda 2^{-3/2}$.

It remains to determine a, given α. To this purpose, we further simplify the analysis by neglecting the probability that a sample path crosses both lines, which in general will be very small. From known results in Brownian motion (see e.g. *Durbin*, 1971, Lemma 3), we now deduce that the probability that a sample path Z_r crosses the line $y = d + c(r\text{-}K)$ for some r ($K \leq r \leq T$) equals

$$Q(\frac{d+c(T\text{-}K)}{(T\text{-}K)^{1/2}}) + \exp(-2dc) \, Q(\frac{d\text{-}c(T\text{-}K)}{(T\text{-}K)^{1/2}}) , \tag{4.21}$$

where $Q(Z) = (2\pi)^{-1/2}\int_Z^\infty \exp(-u^2/2)du$. Substituting $d = a(T\text{-}K)^{1/2}$ and $c = 2a(T\text{-}K)^{-1/2}$ therefore produces the desired equation to be solved for a:

$$Q(3a) + \exp(-4a^2)(1\text{-}Q(a)) = \alpha/2 \tag{4.22}$$

Some useful pairs of values for a and α derived from (4.22) are

$$\begin{aligned}
\alpha &= 1\,\% , \quad a = 1.143 , \\
\alpha &= 5\,\% , \quad a = 0.948 , \\
\alpha &= 10\,\% , \quad a = 0.850 .
\end{aligned} \tag{4.23}$$

Let us however recall the simplifications that have been made along the way to (4.23). First, the W_r's were replaced by the better manageable $\tilde{W}_r$'s. Second, the $\tilde{W}_r$'s were assumed to be discrete observations of a continuous process with a similar covariance structure, and finally, we disregarded the probability that a sample path crosses both critical lines. As a result, the CUSUM test will in general not attain its nominal significance level α in finite samples (a property it shares with many of the procedures discussed so far), and we will for selected X-matrices determine this discrepancy via Monte Carlo experiments.

The CUSUM of squares test, also suggested by *Brown* et al. (1975), relies on successive sums of squares of the recursive residuals. This appears preferable whenever the parameter instability is haphazard rather than systematic, i.e. when positive and negative expected values for the recursive residuals cancel each other out in the CUSUM quantity. Instead one might consider

$$S_r = \Sigma_{t=K+1}^r \tilde{u}_t^2 / \Sigma_{t=K+1}^T \tilde{u}_t^2 \quad (r = K+1,...,T). \tag{4.24}$$

Under H_0, S_r has a beta distribution with mean $(r-K)/(T-K)$, and it seems intuitively reasonable to reject H_0 whenever the S_r's deviate too much from this line, i.e. cross one of the lines

$$g(r) = \pm c + (r-K)/(T-K). \tag{4.25}$$

Similar to the CUSUM test, the crucial step is to determine c, given some significance level α. Again, this is rather complicated and involves some fairly advanced probability theory.

For T-K even, it can be shown (see *Brown et al.*, 1975, p. 154) that the joint distribution of the $\frac{1}{2}$(T-K)-1 statistics $S_{K+2}, S_{K+4},..., S_{T-2}$ is the same as that of an ordered sample of an equal number of independent uniform (0,1) variables. Such situations have been widely studied in mathematical statistics, and there exist tables for the distribution of the maximum deviation from the mean value line (*Pyke*, 1959). Brown et al. suggest to take the resulting significance levels as approximations to the true significance values of

$$c^+ = \max \left(S_{K+r} - \frac{r}{T-K} \right) \quad r = 1,...,T-K-1$$

and

$$c^- = \max \left(\frac{r}{T-K} - S_{K+r} \right) \quad r = 1,...,T-K-1, \tag{4.26}$$

which are the maximum positive and negative deviations of the whole set of S_r's from the mean value line, and again to disregard the probability of crossing both critical lines.

The CUSUM and the CUSUM of squares tests with dummy exogenous variables

In practice. one often has regression models with a constant, where in addition one of the remaining regressor variables is a dummy and also constant for the first T_0 observations. Whenever $T_0 \geq K$, the matrices $X^{(t)}$ (t = K,...,T$_0$) will then have less than full column rank, i.e. the formula (2.16) for the computation of the recursive residuals $\tilde{u}_t$ (t = K+1,...,T$_0$+1) does not apply any more.

$'t = K+1,...,T_0+1)$ does not apply any more.

Here it is important to note that under H_0, the statistical properties of the CUSUM and CUSUM of squares statistics do not depend on the type of LUS residual vector used for the analysis. All we need is the property that under H_0, the $\tilde{u}_t$'s ($t = K+1,...,T$) are NID($0,\sigma^2$). Therefore, when recursive residuals are not available, one can switch to another set of LUS residuals and still use the same tables and significance points. The only problem is to choose these such that the source of the diagnostic power of the recursive residuals (i.e. the fact that u_t depends only on observations prior to t) is preserved as far as possible.

Brown et al. (1975, pp. 152, 153) suggest to proceed as follows:

Drop the constant regressor at the beginning of the recursions, and derive recursive residuals $\tilde{u}_t^0$ ($t = K,...,T_0$) from estimates $\hat{\beta}_0^{(t)}$ ($t = K-1,...,T_0-1$), where β_0 is the $(K-1)\times 1$ subvector of β corresponding to the remaining regressors. Then when the dummy regressor has changed bring it into the regression and calculate the recursive residuals $\tilde{u}_t$ ($t = T_0+2,...,T$) by formula (2.16) as in the standard case. There will be no residual at $t = T_0+1$ (since there is no forecast for y_{T_0+1}), but the overall number of residuals is still T-K. Also, by defining $\tilde{u}_t = \tilde{u}_{t-1}^{(0)}$ for $t = K+1,...,T_0+1$, the index set for the adjusted residuals is the same as in the standard case.

It remains to prove that this procedure produces a LUS residual vector. Since both the $\tilde{u}_t^0$'s ($t = K,...,T_0$) and the $\tilde{u}_t$'s ($t = T_0+2,...,T$) are NID($0,\sigma^2$) when considered separately, what has to be shown is that

$$E(\tilde{u}_s^0\,\tilde{u}_t) = 0 \;(K\leq s\leq T_0,\, T_0+2\leq t\leq T)\,. \tag{4.27}$$

To this purpose, assume without loss of generality that the dummy variable has value zero for $1\leq t\leq T_0$, and is the last variable in the regression. We can therefore, in the notation of section 2.a, for $s\leq T_0$ write the partial regressor matrices $X^{(s)}$ as $X^{(s)} = [\tilde{X}^{(s)}:0]$, and

$$\tilde{u}_s^0 = (y_s - \tilde{x}_s\,'\hat{\beta}_0^{(s-1)})/\tilde{f}_s\,, \tag{4.28}$$

where $\tilde{x}_s$ is the s'th row of X with the last element deleted,

$$\hat{\beta}_0^{(s-1)} = (\tilde{X}^{(s-1)}\,'\tilde{X}^{(s-1)})^{-1}\tilde{X}^{(s-1)}\,'y^{(s-1)} \tag{4.29}$$

is the OLS estimate for the first K-1 coefficients of β based on the first s-1 observations, and

$$\tilde{f}_s = (1+\tilde{x}_s\,'(\tilde{X}^{(s-1)}\,'\tilde{X}^{(s-1)})x_s)^{1/2}\,. \tag{4.30}$$

We note next that for $K\leq s\leq T_0$ and $T_0+2\leq t\leq T,$ $\tilde{u}_s^0$ and $\tilde{u}_t$ can now be expressed as $\tilde{u}_s^0 = \tilde{c}\,'u$ and $\tilde{u}_t = c\,'u$, where

$$\tilde{c}\,' = \tilde{f}_s^{-1} (-\tilde{x}_{s'}(\tilde{X}^{(t-1)}\,'\tilde{X}^{(t-1)})^{-1}\tilde{X}^{(s-1)}\,',1,0,...,0) \tag{4.31}$$

and

$$c\,' = f_t^{-1} (-x_t\,'(X^{(t-1)}\,'X^{(t-1)})^{-1}X^{(t-1)}\,', 1,0,...,0)$$

(see 2.18), so it suffices to show that $\tilde{c}'c=0$. However, straightforward computation produces

$$\tilde{c}'c = \tilde{f}_s^{-1}f_t^{-1}\{\tilde{x}_s'(\tilde{X}^{(s-1)'}\tilde{X}^{(s-1)})^{-1}\tilde{X}^{(s-1)'}\tilde{X}^{(t-1)}(X^{(t-1)'}X^{(t-1)})^{-1}x_t$$
$$- x_s'(X^{(t-1)'}X^{(t-1)})^{-1}x_t\}, \qquad (4.33)$$

and the required result follows immediately from

$$\tilde{X}^{(s-1)'}X^{(t-1)} = [\tilde{X}^{(s-1)'}\tilde{X}^{(s-1)}:0] \qquad (4.34)$$

and the fact that the first term in (4.33) can in view of (4.34) be written as

$$\tilde{x}_s'[I:0](X^{(t-1)'}X^{(t-1)})^{-1}x_t = x_s'(X^{(t-1)'}X^{(t-1)})^{-1}x_t. \qquad (4.35)$$

Alternatives to the CUSUM and CUSUM of squares tests

The power of both the CUSUM and the CUSUM of squares tests depends very much on the timing of possible shifts. For instance, when the parameters change only late in the sample, neither test has much time to pick this up and power will be rather low. One way to avoid this, already suggested by *Brown et al.* (1975), is to reverse the order of observations, i.e. to run the recursive estimation process backwards. A related test by *Schweder* (1976) still uses the recursive residuals as computed in the forward manner and only does the summing up backwards. There is some evidence that this is preferable whenever the regression relationship remains constant for more than K initial periods. If a single shift occurs at T^*, where $T^* > K+1$, the first T^*-K-1 recursive residuals contribute only noise to the forward CUSUM (similarly to the forward CUSUM of squares), which does not happen when the summation is done backwards. It is essential that forward residuals are used for the backward CUSUMs, since otherwise the initial recursive residuals would again only contribute noise to the test statistic.

A related procedure, due to *Bauer* and *Hackl* (1978) and *Hackl* (1980), is based on moving rather than cumulated sums of recursive residuals. Their MOSUM test relies on the quantities

$$M_t = \frac{1}{s}\Sigma_{i=t-G+1}^{t}\tilde{u}_i \quad (t=K+G,...,T), \qquad (4.36)$$

where G is a fixed number of terms included in each moving sum (MOSUM), and s is the familiar estimate for the standard deviation σ of the disturbances. Since there is always a fixed number of terms in M_t, the relative importance of recursive residuals that due to a zero mean do not contribute to a significant test statistic is automatically limited. There is however no general rule for choosing G, and the distribution under H_0 of $\max_{K+G\leq t\leq T}M_t$ appears rather unmanageable.

The same holds for the MOSUM of squares tests, which is based on the quantities

$$MS_t = \Sigma_{i=t-G+1}^{t}\, \tilde{u}_i^2 / \Sigma_{i=K+1}^{T}\, \tilde{u}_i^2 \quad (t = K+G,...,T). \tag{4.37}$$

Under H_0, the MS_t's have a Beta $B(G/2,(T-K-G)/2)$ distribution, with constant mean value $E(MS_t) = G/(T-K)$. Again, however, it is hard to evaluate whether or not the maximum departure of the M_t's from this mean value line is significant. In particular, the arguments that Brown et al. have used to evaluate the null distribution of the CUSUM of squares statistics (4.24) do not apply here.

McCabe and *Harrison* (1980) have suggested to stay with the CUSUM of squares statistic as defined in (4.24), but with OLS rather than recursive residuals. The main advantage of this procedure is its computational simplicity. Since however the distribution of the OLS residual vector $\hat{u}$, and therefore also the distribution of the OLS-based CUSUM of squares statistics $S_r (r = K+1,...,T)$ depend on the regressor matrix X, the null distribution of the S_r's is now much harder to evaluate. McCabe and Harrison suggest bounding random variables S_{rL} and S_{ru}, whose distributions do not depend on X, such that $S_{rL} \leq S_r \leq S_{ru}$ for all r $(K+1 \leq r \leq T)$.

There are now four rather than two critical lines of the form (4.25) for any significance level α, with rejection resulting from crossing either (or both) of the outer lines. The null hypothesis of parameter constancy is accepted when the sample CUSUM of squares plot stays within the inner lines, and the test is inconclusive when the plot crosses the inner lines only. This inconvenience, which the procedure shares with the Durbin-Watson test or the Harrison-McCabe test for heteroskedasticity, has prohibited a widespread acceptance in empirical work.

Following *Brown et al.* (1975), most procedures discussed in this subsection have been developed without regard to any particular parameterized alternative hypothesis. Once such an alternative is specified, any one of the Wald-, Lagrange Multiplier or Likelihood Ratio principles can be employed to derive yet another test. Proceeding along these lines, *Garbade* (1977) has investigated the alternative that the β_t's in (4.10) are stochastic and follow a random walk with zero drift through time:

$$\beta_t = \beta_{t-1} + p_t, \quad (t = 2,...,T), \tag{4.38}$$

where the p_t's are NID$(0,P)$ and independent of the regression disturbances u_t. Such variable parameter models are discussed at length in the Kalman Filter literature (see e.g. *Athans*, 1974, or *Cooley* and *Prescott*, 1976). The null hypothesis of parameter constancy is here obviously equivalent to H_0: $P = 0$. Garbade suggests a likelihood ratio procedure to test whether or not this is true. Again, the null distribution of the test statistic is hard to evaluate and is in particular not well approximated by a χ^2 distribution in small samples.

There are also various procedures that explicitly rely on parameter estimates as indicators of possible structural shifts. *Brown et al.* (1975) have proposed to fit a regression to G successive observations $(G \geq K)$ and to move this segment along the series, as a supplement to their CUSUM and CUSUM of squares tests. The graphs of the resulting estimates of the elements of β are then supposed to provide further evidence of departure from constancy (though not a formal test). The disturbance variance σ^2 may also be estimated separately for each data segment and plotted against time.

Dufour (1982) has suggested a direct parameter based analogue of the CUSUM test. His point of departure is the relationship

$$\hat{\beta}^{(t)} = \hat{\beta}^{(t-1)} + (X^{(t)}\,'X^{(t)})^{-1}\,x_t(y_t - x_t\,'\hat{\beta}^{(t-1)}) , \tag{4.39}$$

which follows immediately from

$$
\begin{aligned}
X^{(t)}\,'X^{(t)}\hat{\beta}^{(t)} &= X^{(t)}\,'y^{(t)} \\
&= X^{(t-1)}\,'y^{(t-1)} + x_t y_t \\
&= X^{(t-1)}\,'X^{(t-1)}\hat{\beta}^{(t-1)} + x_t y_t \\
&= X^{(t)}\,'X^{(t)}\hat{\beta}^{(t-1)} + x_t y_t - x_t x_t\,'\hat{\beta}^{(t-1)} .
\end{aligned}
\tag{4.40}
$$

The basic equation (4.39) immediately produces a link between recursive parameter estimates and recursive residuals via

$$
\begin{aligned}
\hat{\beta}^{(t)} - \hat{\beta}^{(t-1)} &= (X^{(t)}\,'X^{(t)})^{-1}\,x_t(y_t - x_t\,'\hat{\beta}^{(t-1)}) \\
&= f_t(X^{(t)}\,'X^{(t)})^{-1}\,x_t \tilde{u}_t, \quad (t = K+1,\ldots,T).
\end{aligned}
\tag{4.41}
$$

This suggests a test based on standardized *first differences* of recursive parameter estimates. Under H_0, the changes in the parameter estimates as one proceeds with the recursive process are independent and normal with mean

$$E(\hat{\beta}^{(t)} - \hat{\beta}^{(t-1)}) = 0 \quad (t = k+1,\ldots T) \tag{4.42}$$

and covariance matrices

$$\operatorname{cov}(\hat{\beta}^{(t)} - \hat{\beta}^{(t-1)}) = \sigma^2 f_t^2\,(X^{(t)}\,'X^{(t)})^{-1}x_t x_t\,'(X^{(t)}\,'X^{(t)})^{-1} . \tag{4.43}$$

Following Dufour (1982), let $a_j^{(t)}$ denote the j'th column of $(X^{(t)}\,'X^{(t)})^{-1}$. The j'th component of $\hat{\beta}^{(t)} - \hat{\beta}^{(t-1)}$ $(1 \leq j \leq K)$ can then be written as

$$
\begin{aligned}
\hat{\beta}_j^{(t)} - \hat{\beta}_j^{(t-1)} &= f_t(a_j^{(t)}\,'x_t)\tilde{u}_t \\
&= D_j^{(t)}\tilde{u}_t ,
\end{aligned}
\tag{4.44}
$$

where $D_j^{(t)} = f_t(a_j^{(t)}\,'x_t)$. If $D_j^{(t)} \neq 0$ $(j=1,\ldots,K; t=K+1,\ldots,T)$ the resulting standardized differences

$$\Delta_j^{(t)} = (\hat{\beta}^{(t)} - \hat{\beta}_j^{(t-1)})/|D_j^{(t)}| \tag{4.45}$$

are then $NID(0,\sigma^2)$ under H_0. The K-vectors

$$\Delta_j = (\Delta_j^{(K+1)},\ldots,\Delta_j^{(T)})' \quad (j=1,\ldots,K) \tag{4.46}$$

constitute K sets of LUS residuals, which in view of (4.44) are linked to the recursive residuals via

$$\Delta_j^{(t)} = (D_j^{(t)}/|D_j^{(t)}|)\tilde{u}_t \quad (t = K+1,\ldots,T) . \tag{4.47}$$

The elements of $\Delta_j^{(t)}$ have thus the same absolute values as $\tilde{u}_t$. *They may however ex hibit very different sign patterns*, and *Dufour* (1982) provides some evidence that the first differences of parameter estimates can be much more revealing concerning structural change than the recursive residuals.

The Fluctuation test

Finally we present a procedure which *Ploberger* (1983) and *Kontrus* and *Ploberger* (1984) have called the "Fluctuation Test". This name derives from the rule to reject the null hypothesis of parameter constancy whenever the recursive parameter estimates $\hat{\beta}^{(t)}$ $(t = K+1,\ldots,T)$ fluctuate too much. Contrary to Dufour's procedure, the Fluctuation test is thus based on the *levels*, rather than the first differences, of successive recursive parameter estimates. A similar rule has also been suggested by *Sen* (1980), but only for single regressor models and under severe restrictions on the independent variable.

The Dufour procedure rejects the null hypothesis whenever the first differences of the successive parameter estimates or their cumulated sums deviate too much from zero. Unfortunately, the analoguous natural yardstick against which to compare the fluctuations of the parameter estimates themselves, i.e. the true parameter vector β, is not known. Ploberger and Kontrus replace it by the full sample estimate $\hat{\beta}^{(T)}$, and base the test on the fluctuations, as $t = K,\ldots,T$, of

$$||\hat{\beta}^{(t)} - \hat{\beta}^{(T)}||_\infty = \max_{i=1,\ldots,K} |\hat{\beta}_i^{(t)} - \hat{\beta}_i^{(T)}|. \tag{4.48}$$

The test statistic is

$$F^{(T)} = \max_{t=K,\ldots,T} ||F_t^{(T)}||_\infty , \tag{4.49}$$

where

$$F_t^{(T)} = \frac{t-K}{S(T-K)} (X^{(T)'}X^{(T)})^{1/2} (\hat{\beta}^{(t)} - \hat{\beta}^{(T)}),$$

and the null hypothesis is rejected whenever $F^{(T)}$ is too large.

Figure 4.3 shows a typical sample trajectory of the $F_t^{(T)}$'s. Note that these always pass through the points $(K,0)$ and $(T,0)$, and are nonnegative otherwise. Contrary to the CUSUM test, the location of the point T^* where the sample trajectory first crosses the critical line does here not allow any inference concerning the timing of a possible structural change, since such a shift affects both $\hat{\beta}^{(t)}$ and $\hat{\beta}^{(T)}$.

The particular norm in (4.48) is arbitrary. Similar to the other ingredients in (4.49), it is mainly used to facilitate the evaluation of the null distribution of the test statistic (4.49).

FIG.4.3: SAMPLE TRAJECTORY FOR FLUCTUATION TEST

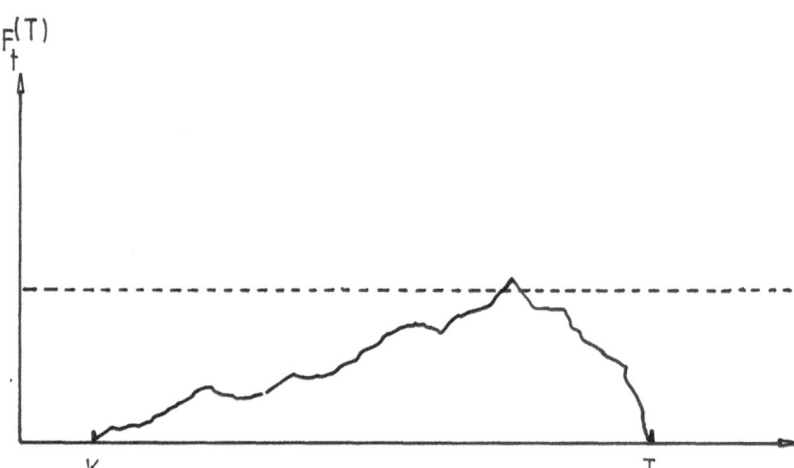

The finite sample null distribution of $F^{(T)}$ is unknown, so the Fluctuation test is only an asymptotic test. Given that the regressors are bounded (note that this excludes trended data), *Ploberger* (1983) shows that under H_0 ,

$$P(F^{(T)} \leq c) \rightarrow (1-f(c))^K \qquad (4.50)$$

as $T \rightarrow \infty$, where $c > 0$ and where

$$f(c) = 2 \sum_{i=1}^{\infty} (-1)^{i+1} \exp(2i^2 c^2) \qquad (4.51)$$

The proof of this result is rather lengthy and relies heavily on the asymptotic distribution theory of probability measures on metric spaces, so we only sketch it here.

The problem is: given some significance level α and the corresponding critical value c_α for the test statistic $F^{(T)}$, how can we derive

$$\lim_{T \rightarrow \infty} P(F^{(T)} > c_\alpha)$$

from the corresponding probability of a suitably defined limit process? Unfortunately finite dimensional distribution theory does not apply here, since the processes $F_t^{(T)}$ (where there is one process for any given T, indexed by the integers $t = K,...,T$) do not converge in distribution (in the ordinary sense) to anything.

The common way out of this dilemma is to normalize these processes so that each one is indexed by the real numbers in the [0,1] interval. This is achieved by defining, for $0 \leq z \leq 1$,

$$F^{(T)}(z) = F^{(T)}_{[K+(T-K)z]},$$

(4.52)

where $[K+(T-K)z]$ is the floor of $K+(T-K)z$, i.e. the largest integer less than or equal to $K+(T-K)z$. These processes $F^{(T)}(z)$ can also be viewed as mappings from the underlying probability space in a suitably defined space of functions, and are then called random elements.

The function space can be endowed with a metric, and the theory of convergence in distribution of random elements can then be treated as a special case of the weak convergence of probability measures on metric spaces, as discussed in e.g. *Billingsley* (1968). *Breiman* (1968, chapters 12 and 13) and *Gänssler* and *Stute* (1977) also provide more elementary textbook introductions to the rather complex issues that are involved here. A convenient summary of the state of the art can also be found in *Serfling* (1980, chapter 1.11).

In the present context, the basic idea is to show that the random elements

$$\frac{K+(T-K)z}{S\sqrt{T-K}} Q^{1/2} \left(\hat{\beta}^{[K+(T-K)z]} - \beta \right)$$

(4.53)

converge in distribution to a K-dimensional Wiener Process $W(z)$. From this, one can establish that the random elements $F^{(T)}(z)$ converge in distribution to the K-dimensional version of a process known as the Brownian Bridge (or tied-down Brownian Motion; see Billingsley, 1968, p. 65).

The limiting process $F(z)$ is related to $W(z)$ by $F(z) = W(z) - z \, W(1)$, so $F(0) = F(1) = 0$.

From the well established distribution theory of the Brownian Bridge, one can then immediately deduce (4.50). The critical values F_α such that

$$P(\sup_{0 \leq z \leq 1} \| F(z) \|_\infty > F_\alpha) = \alpha$$

are for some K and α reproduced in table 4.1.

Table 4.1: BOUNDARIES FOR FLUCTUATION TEST

Significance level	1	2	3	4	5	6	7	8	9	10
1%	1.63	1.73	1.79	1.83	1.86	1.88	1.90	1.92	1.94	1.95
5%	1.36	1.48	1.54	1.59	1.62	1.65	1.67	1.69	1.71	1.73
10%	1.22	1.35	1.42	1.47	1.51	1.54	1.57	1.59	1.61	1.62

Local power

We next discuss the relative power of the CUSUM, CUSUM of squares and Fluctuation tests, confining ourselves first to the asymptotic case. A Monte Carlo examination of the finite sample case will follow in the next subsection.

Again, any rigorous discussion involves a fair amount of probability theory on metric spaces, so we will only sketch the main results. A more detailed discussion (with proofs) can be found in *Ploberger* (1983) and *Ploberger* and *Krämer* (1986 b).

When examining the local power of the tests, the first problem is to find a suitable sequence of alternatives such that the intensity of the structural change decreases as sample size increases. We proceed as follows: Let $g(z)$ be a K-valued step function (or a function that can be uniformly approximated by a step function) on the $[0,1]$ interval. A sequence of local alternatives is then given by the following triangle scheme of regression coefficients (M any real number):

$$\beta_t^{(T)} = \beta + \frac{M}{\sqrt{T}} \, g(\tfrac{t}{T}) \, . \tag{4.54}$$

If for instance $g(z) = 0$ for $z < z_0$ and $g(z) = \Delta\beta$ for $z \geq z_0$ $(0 < z_0 < 1)$, this corresponds to a single shift in the regression parameters at time $T^* = z_0 T$, i.e. at the z_0-quantile of the observations. The point of this shift moves along the time axis in proportion to the sample size, and its intensity decreases with $1/\sqrt{T}$ as $T \to \infty$.

For sequences of alternatives given by (4.54), Ploberger (1983) and Ploberger and Krämer (1986b) show the following:

(i)   Let $p^{(T)}(M)$ be the rejection probability of the Fluctuation test corresponding to some $M > 0$ and some significance level $\alpha > 0$, and assume that there does not exist a subset of the $[0,1]$ interval with Lesbesque measure one on which $g(z)$ is constant. Then

$$\lim_{M \to \infty} \lim_{T \to \infty} p^{(T)} (M) = 1 \, . \tag{4.55}$$

(ii)   Let $\tilde{p}^{(T)}(M)$ be the corresponding rejection probability of the CUSUM of squares test. Then

$$\lim_{T \to \infty} \tilde{p}^{(T)} (M) = \alpha \tag{4.56}$$

irrespective of $g(t)$ (i.e. the CUSUM of squares test has only trivial local power).

(iii)   Let $p^{*(T)}(M)$ be the rejection probability of the CUSUM test, and assume that the "mean regressor"

$$q = \lim_{T \to \infty} \frac{1}{T} \Sigma_{t=1}^{T} x_t$$

exists (this will always hold in view of 2.3 when there is a constant in the regression, in which case q will equal the first column of Q). Then, if q is orthogonal to $g(z)$ for all z, the equivalent of (4.56) holds:

$$\lim_{T \to \infty} p^{*(T)}(M) = \alpha \, . \tag{4.57}$$

These results have important implications. First, they establish the Fluctuation test as the superior procedure among those considered here (given that the concept of local power is the appropriate efficiency criterion; later our Monte Carlo experiments will also produce some finite sample evidence for this). This is the onlv test whose local power can be made arbitrarily large irrespective of the function g(z). Second, and somewhat unexpectedly, they favour the CUSUM test as compared to the CUSUM of squares test, whereas conventional wisdom implies the reverse. The latter has only trivial local power, no matter what type of structural change occurs, whereas the local power of the former is trivial only when the structural change is orthogonal to the mean regressor q. Otherwise, the local power of the CUSUM test will in general exeed the significance level α, and the CUSUM test will in this sense outperform the CUSUM of squares test.

The popular distrust of the CUSUM test as echoed in e.g. *McCabe* and *Harrison* (1980) or *Johnston* (1984,p.392), probably originates in the devastating Monte Carlo results in *Garbade* (1977), where the empirical power of the CUSUM test is extremely low. However, Garbade had $K = 1$ and NID $(0,25)$ regressors in his experiments. This implies $q = 0$, i.e. *the mean regressor is orthogonal to any type of structural change* . Not surprisingly, therefore, the CUSUM test did not perform particularly well.

FIG.4.4: STRUCTURAL SHIFT ORTHOGONAL TO THE MEAN REGRESSOR

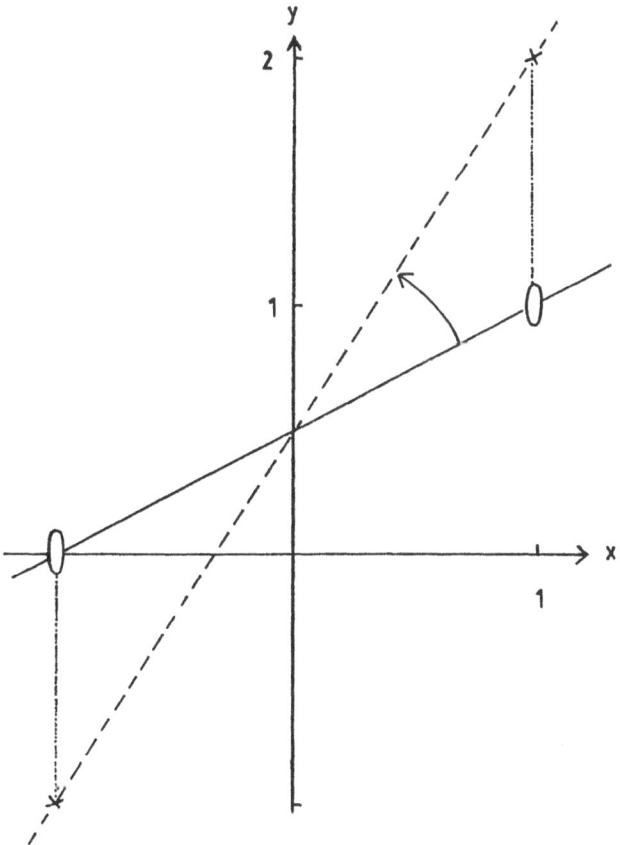

Figure (4.4) illustrates this important point. For simplicity, we consider the regression

$$y_t = \frac{1}{2}(-1)^t + \frac{1}{2} + u_t,$$

with a one-time increase in slope from $1/2$ to $3/2$ at time T^*. The parameter shift $\Delta\beta = [1,0]$ is thus orthogonal to the mean regressor $q = [0,1/2]$. If $x_{T^*+1} = 1$, the forecast for y_{T^*+1}, being derived from the old regression line, will be $\hat{y}_{T^*+1} = 1$, producing a forecast error $y_{T^*+1} - \hat{y}_{T^*+1} = 1$ (neglecting any stochastic variability). Similarly, the forecast for y_{T^*+2} will still be dominated by the old regression line and equal to $\hat{y}_{T^*+2} = 0$, leading to a forecast error of -1. Only gradually will the estimated regression line approximate the new true regression, and will the systematic forecast errors decrease in absolute value. They will continue however to have opposite signs, so the CUSUMs have no chance to cumulate, and the power of the CUSUM test to detect this type of parameter shift will be very low.

A Monte Carlo comparison of the CUSUM and Fluctuation tests

Next we evaluate the finite sample performance of the CUSUM and Fluctuation tests via some Monte Carlo experiments. The CUSUM of squares test is not considered, since it has already been shown to be (asymptotically) inferior.

To save computer time, we confine ourselves to the case $K=2$ (a bivariate regression with a constant), where

$$x_t = [1,(-1)^t]' \qquad (t = 1,...,T) . \tag{4.58}$$

The particular independent variable was chosen to ensure that (2.3) holds, and that the mean regressor q exists, where

$$q = \lim_{T \to \infty} \frac{1}{T} \Sigma_{t=1}^{T} x_t = [1,0]' . \tag{4.59}$$

Since both the CUSUM and Fluctuation tests attain their nominal significance level only asymptotically, we first determined the actual significance levels for various values of T and $\alpha = 5\%$. The disturbances were generated as NID(0,1) variates, using various NAG-library subroutines, and $\beta = [0,0]'$ was used throughout.

The results of the simulations are reproduced in table (4.2). N = 10.000 independent runs were performed for each value of T. Given a Monte Carlo estimate $\hat{\alpha}$ for the true α, the standard error of $\hat{\alpha}$ was computed from

$$S = (\hat{\alpha}(1-\hat{\alpha})/N)^{1/2}. \tag{4.60}$$

These values are shown in parentheses below the α-figures.

Table 4.2: TRUE REJECTION PROBABILITIES UNDER H_0 FOR $\alpha = 5\%$

	T			
	30	60	120	1000
CUSUM test	2.13	3.26	3.30	4.20
	(.14)	(.18)	(.18)	(.20)
Fluctuation test	.29	1.22	2.30	4.10
	(0.5	(.11)	(.15)	(.20)

The table demonstrates that the actual significance levels are for both tests consistently below the nominal ones and quite dramatically so for the Fluctuation test and small T, which is in line with *Anderson* (1975) and the theoretical arguments in *Ploberger* (1983). The difference however decreases as T increases, as predicted by the asymptotic theory.

Our investigation of the power of the tests is confined to the simplest case of a single shift in the regression parameters, mainly for ease of comparison with previous work (e.g. *Garbade*, 1977; *Hackl*, 1980; *McCabe* and *Harrison*, 1980; *Kontrus* and *Ploberger*, 1984). We keep the basic model as before and confine ourselves to the empirically most relevant sample sizes T = 30, 60, 120. Also, the number of replications per parameter combination is now 1000 rather than 10.000 to save computer time.

We systematically vary (i) the timing, (ii) the intensity, and (iii) the angle of the structural shift. The shift occurs at time $T^* = dT$, where d takes the values 0.3, 0.5 and 0.7, respectively. This corresponds to a shift early, midway and late in the sample period. The shift itself is given by

$$\Delta \beta = \frac{b}{\sqrt{T}}[\cos \Psi, \sin \Psi], \tag{4.61}$$

where Ψ is the angle between $\Delta\beta$ and the mean regressor $q = [1,0]'$. Similar to our investigation of the finite sample null distribution, β is initially set to zero. Thus the parameter vector β equals $\beta = [0,0]'$ for t up to T^* and $\Delta\beta$ subsequently.

The intensity of the shift is $||\Delta\beta|| = |b|/\sqrt{T}$. We let this go down with increasing sample size so that the rejection probabilities for d, b and Ψ approximate the local power of the tests as T increases.

Table (4.3) reports the results of our experiments, for $\alpha = 5\%$, various values of b, d and Ψ, and $T = 30$. The results for $T = 60$ and $T = 120$ are in tables (4.4) and (4.5), respectively.

Most strikingly, the performance of the CUSUM test indeed deteriorates as Ψ increases, confirming the theoretical results above. When the structural shift is orthogonal to the mean regressor ($\Psi = 90°$), the power is even less than the nominal significance level of $\alpha = 5\%$.

By contrast, the Fluctuation test is only marginally affected by the angle Ψ. Ceteris paribus, its power is lowest when Ψ is around $45°$, a consequence of the norm (4.48). For a given intensity of the shift, as measured in the Euclidean norm, the norm (4.48) is smallest when $\Psi = 45°$.

Ceteris paribus, an early shift favours the CUSUM test, whose power then declines as the shift point moves towards the end of the sample. This is in line with both a priori expectations (since the CUSUMs have no chance to cumulate when the shift occurs too late) and earlier Monte Carlo experiments (e.g. *Hackl*, 1980, chapter 7). On the other hand, the Fluctuation test fares best when the shift occurs in the middle of the sample.

Both tests perform better as the intensity of the structural shift increases, except for high values of Ψ, where the CUSUM test deteriorates as b increases.

The relative performance of the tests is such that when the shift occurs early in the sample ($d = 0.3$) and Ψ is not too large, the CUSUM test outperforms the Fluctuation test. For most parameter combinations, however, the latter test has greater power, with a margin that is at times substantial.

This superiority of the Fluctuation test appears even more striking in view of its smaller actual size, as shown in table (4.2). If there were a correction for this, the relative performance of the Fluctuation test would be even better.

Thus our experiments, though based on a very special design, nevertheless establish the Fluctuation test as a serious competitor among procedures to detect a structural shift in a regression.

Table 4.3: EMPIRICAL POWER FOR T = 30

	b			angle ψ			
		$0°$	$16°$	$36°$	$54°$	$72°$	$90°$

a) CUSUM Test

	b	$0°$	$16°$	$36°$	$54°$	$72°$	$90°$
d=0.3	4.8	.18	.13	.09	.04	.02	.01
	7.2	.34	.26	.16	.06	.01	.00
	9.6	.58	.53	.27	.07	.02	.00
	12.0	.75	.79	.39	.09	.01	.00
d=0.5	4.8	.08	.07	.05	.02	.01	.01
	7.2	.19	.15	.09	.03	.01	.00
	9.6	.36	.26	.11	.02	.00	.00
	12.0	.50	.41	.17	.02	.00	.00
d=0.7	4.8	.04	.03	.02	.01	.01	.01
	7.2	.05	.04	.02	.01	.00	.00
	9.6	.08	.06	.02	.00	.00	.00
	12.0	.12	.08	.04	.01	.00	.00

b) Fluctuation Test

	b	$0°$	$16°$	$36°$	$54°$	$72°$	$90°$
d=0.3	4.8	.03	.02	.01	.01	.02	.02
	7.2	.08	.05	.02	.02	.05	.06
	9.6	.20	.14	.06	.05	.16	.20
	12.0	.46	.30	.10	.09	.32	.47
d=0.5	4.8	.10	.07	.06	.04	.09	.08
	7.2	.39	.26	.12	.14	.28	.39
	9.6	.77	.62	.29	.27	.59	.72
	12.0	.95	.87	.49	.49	.85	.95
d=0.7	4.8	.06	.05	.01	.02	.03	.05
	7.2	.20	.12	.05	.05	.14	.20
	9.6	.52	.37	.12	.13	.38	.54
	12.0	.81	.68	.23	.23	.63	.82

Table 4.4: EMPIRICAL POWER FOR T = 60

	b	angle ψ					
		$0°$	$16°$	$36°$	$54°$	$72°$	$90°$
a) CUSUM Test							
d=0.3	4.8	.19	.18	.13	.08	.04	.03
	7.2	.42	.41	.26	.11	.04	.01
	9.6	.73	.66	.46	.20	.03	.01
	12.0	.90	.86	.63	.28	.03	.01
d=0.5	4.8	.12	.14	.08	.05	.02	.02
	7.2	.31	.27	.18	.07	.03	.01
	9.6	.56	.50	.31	.11	.02	.01
	12.0	.79	.68	.48	.14	.02	.00
d=0.7	4.8	.06	.06	.05	.04	.03	.02
	7.2	.09	.09	.06	.03	.02	.01
	9.6	.19	.16	.11	.04	.01	.01
	12.0	.34	.26	.14	.05	.01	.00
b) Fluctuation Test							
d=0.3	4.8	.14	.12	.08	.08	.13	.11
	7.2	.37	.34	.21	.21	.32	.40
	9.6	.73	.66	.46	.46	.68	.78
	12.0	.96	.92	.72	.74	.90	.95
d=0.5	4.8	.28	.25	.18	.19	.25	.27
	7.2	.70	.59	.47	.47	.62	.70
	9.6	.95	.90	.80	.79	.91	.95
	12.0	1.00	.99	.95	.96	.98	1.00
d=0.7	4.8	.18	.16	.13	.09	.15	.17
	7.2	.50	.41	.27	.31	.41	.49
	9.6	.84	.75	.59	.60	.74	.82
	12.0	.98	.95	.84	.84	.95	.97

Table 4.5: EMPIRICAL POWER FOR T = 120

	b	\(0°\)	\(16°\)	\(36°\)	\(54°\)	\(72°\)	\(90°\)
				angle ψ			

a) CUSUM Test

	b	0°	16°	36°	54°	72°	90°
d=0.3	4.8	.25	.19	.15	.07	.04	.03
	7.2	.52	.45	.35	.16	.06	.03
	9.6	.79	.75	.55	.26	.08	.02
	12.0	.94	.91	.76	.43	.08	.02
d=0.5	4.8	.17	.14	.10	.06	.04	.03
	7.2	.37	.33	.23	.11	.05	.02
	9.6	.66	.60	.42	.18	.05	.02
	12.0	.87	.80	.71	.27	.05	.01
d=0.7	4.8	.07	.07	.05	.05	.04	.02
	7.2	.13	.13	.11	.06	.04	.03
	9.6	.28	.23	.14	.08	.02	.02
	12.0	.47	.41	.26	.09	.03	.01

b) Fluctuation Test

	b	0°	16°	36°	54°	72°	90°
d=0.3	4.8	.22	.20	.16	.17	.20	.20
	7.2	.58	.51	.41	.41	.49	.58
	9.6	.88	.86	.72	.72	.83	.88
	12.0	.99	.97	.92	.92	.97	.99
d=0.5	4.8	.37	.33	.28	.28	.32	.37
	7.2	.79	.74	.66	.65	.73	.79
	9.6	.97	.96	.91	.91	.96	.97
	12.0	1.00	1.00	.99	.99	1.00	1.00
d=0.7	4.8	.23	.23	.19	.19	.23	.24
	7.2	.60	.57	.47	.46	.56	.62
	9.6	.90	.85	.76	.78	.86	.90
	12.0	.99	.98	.95	.94	.98	.99

Dynamic models

All tests in this chapter require that the assumptions of the standard linear regression model hold under the null hypothesis (possibly excluding normality). In particular, the regressors are assumed nonstochastic, which excludes lagged endogenous variables. Of course this is a major drawback. The issue of whether or not the tests generalize to dynamic models is thus of considerable practical importance. Below, we briefly address this problem for some CUSUM type procedures.

For simplicity, consider the simple dynamic model

$$y_t = \gamma \, y_{t-1} + \beta_1 \, x_{t1} + \ldots + \beta_K \, x_{tK} + u_t \qquad (t = 1, \ldots, T). \tag{4.62}$$

Dufour (1982, p.46) has noted that if we knew the true value of γ in (4.62), the model could be reduced to standard form by considering

$$y_t - \gamma \, y_{t-1} = \beta_1 \, x_{t1} + \ldots + \beta_K \, x_{tK} + u_t. \tag{4.63}$$

One could then proceed as usual and recursively estimate the vector β. When γ is unknown, one can replace it by the OLS estimate $\hat{\gamma}$ from the full sample, and hope that the resulting recursive residuals and any tests based on them will have approximately the same properties as those based on the true γ. We call this the Dufour test.

Krämer et al. (1985) have shown that the Dufour test indeed attains its correct size under H_0 (asymptotically).

The requirements are that $|\gamma| < 1$, and that in addition to (2.3), the lagged cross moments of the regressors also tend to finite limit matrices and that there is a constant in the regression. The disturbances u_t however need only be IID $(0, \sigma^2)$ (not necessarily normal).

The proof again requires some nonelementary probability theory, so we give only the main idea:

Assume that y_0 is known, let

$$y_t^* = y_t - \hat{\gamma} \, y_{t-1} \qquad (t = 1, \ldots, T), \tag{4.64}$$

and compute recursive residuals and the CUSUM quantities (4.9) from the model

$$y_t^* = \beta_1 \, x_{t1} + \ldots + \beta_K \, x_{tK} + u_t^*, \tag{4.65}$$

where $u_t^* = u_t + (\gamma - \hat{\gamma}) y_{t-1}$. We still symbolize recursive residuals and CUSUM quantities by $\tilde{u}_t$ and W_r, respectively, although the recursive residuals from (4.65) now also vary with T (since the dependent variables y_t^* also depend on T via $\hat{\gamma}$). They are of course neither normal nor independent any more, but we will see that this does not matter asymptotically.

The CUSUM quantities W_r cross the critical lines (4.20) if and only if

$$\max_{r = K+1, \ldots, T} \left| \frac{W_r}{\sqrt{T-K}} \right| / \left(1 + 2 \, \frac{r-K}{T-K} \right) > a, \tag{4.66}$$

where the constant a depends on the significance level α and is for α equal to one, five and ten percent given in (4.23). What needs to be shown, therefore, is that under H_0,

$$\lim_{T \to \infty} P(\max_{\tau = K+1,\ldots,T} \left| \frac{W_r}{\sqrt{T-K}} \right| /(1 + 2\frac{r-K}{T-K}) > a) = \alpha . \tag{4.67}$$

This is difficult enough in the standard case without lagged dependent variables and still more involved in the present context. Similar to the derivation of the asymptotic null distribution of the test statistic for the Fluctuation test, it is done by considering the standardized random element

$$W^{(T)}(z) = \frac{1}{\sqrt{T-K}} W_{\tau(z)} , \tag{4.68}$$

where $\tau(z) = [K + (T-K)z]$ is shorthand for the floor of $K + (T-K)z$.

Obviously, (4.66) holds if and only if

$$\max_{0 < z \leq 1} |W^{(T)}(z)|/(1 + 2z) > a . \tag{4.69}$$

The trajectories of the process $W^{(T)}(z)$ are constant on the half open intervals $[(n-1)/(T-K), n/(T-K)]$ ($n = 1,\ldots,T-K$), so $W^{(T)}(z)$ is a random element in the space $D[0,1]$ of all real valued functions on the unit interval that are right continuous and have left limits. Also among the random elements of this space, if endowed with the so called Skorohod metric (see Billingsley, 1968, chapter 3), is the familiar Wiener process $W(z)$. Therefore it suffices to show that the random elements $W^{(T)}$ tend in distribution to W:

$$W^{(T)} \overset{d}{\to} W. \tag{4.70}$$

This is here defined by

$$P(W^{(T)} \epsilon H) \to P(W \epsilon H)$$

for all measurable subsets of $D[0,1]$ with boundary of W-measure zero, which generalizes the usual convergence concept for probability distributions to metric spaces. Since it is easily established that the boundary of the event

$$\max_{0 < z \leq 1} |W(z)|/(1 + 2z) > a \tag{4.71}$$

has W-measure zero, the desired result then follows from (4.70).

The hard part, which we omit here, is to show (4.70). It is here that the requirements listed above come into play, with details given in Krämer et al. (1985).

Although the asymptotic null distribution of the Dufour test does not cause problems in view of the above result, there remain doubts about a possible loss of power induced by the estimation of γ. First, the procedure is unlikely to detect changes in γ when the remaining parameters remain constant. At least, the usual rationale behind the CUSUM test does not apply here. Second, also when γ is constant and only the β's change, a loss of power can occur through the resulting distortion of $\hat{\gamma}$. A more informative procedure would certainly be to apply the CUSUM test to the untransformed model (4.62) without first eliminating the dynamics. We call this the dynamic CUSUM test. Krämer et al. (1985) show that the equivalent of (4.67) holds for this test as well, i.e. that the CUSUM test has asymptotically the correct size whether or not there is a lagged dependent variable in the equation. The proof is again by establishing (4.70).

In view of these results, any choice between the Dufour test and the dynamic CUSUM test is only a matter of power, which we examine next via some Monte Carlo experiments. The experimental design is similar to the one used before, i.e. $x_t = [1,(-1)^t]'$ and $u_t \sim \mathrm{NID}(0,1)$. In addition, we set $\gamma = 0.5$ and $y_0 = 0$.   1.000 replications were performed for any given parameter combination.

We let the structural shift affect only the β-parameters, in order not to challenge the Dufour test outside its home turf, so to speak. The shift again occurs at time $T^* = dT$, where d takes the values 0.3, 0.5 and 0.7, respectively, and the shift itself is again given by (4.61). Initially, $\beta = [1,1]'$. Since there is no shift in γ, the shift in the complete parameter vector $\delta = [0.5, 1, 1]$ is

$$\Delta\delta = \frac{b}{\sqrt{T}} [0, \cos \Psi, \sin \Psi] ,$$

so Ψ is again the angle between $\Delta\delta$ and the mean regressor of the dynamic model (which for $|\gamma| < 1$ always exists).

Table (4.6) summarizes the experiments for $\alpha = 5\%$, various values of b, d, Ψ, and $T = 120$. We do not explicitly reproduce the results for $T = 30$ and $T = 60$, since they are basically similar, exept that the power of all tests is lower. Under the heading "Static CUSUM test" we report the results for γ known.

In table (4.6), several regularities stand out: (i) confirming the results in tables (4.3) - (4.5), the power of the test very much depends on the angle Ψ, and is very low when the structural shift is orthogonal to the mean regressor, (ii) the power increases for all tests with increasing intensity of the shift, except for the high values of Ψ, (iii) the power decreases with increasing d, again confirming our previous results, (iv) the dynamic CUSUM test outperforms the Dufour test, and (v) none of the feasible tests attains the power of the static CUSUM test.

These results also repeated themselves for alternative values of α, β, γ, Ψ, X and T. In particular, it is not surprising that the power of the tests will ceteris paribus decrease with increasing d, since a shift late in the sample period means that neither test has much time to pick this up. Unfortunately, the usual remedy from the static model, i.e. to do the CUSUM test backwards to counterbalance this deficiency, does not apply here, since the initial ordering of the observations is essential to establish the asymptotic null distribution of the test statistics.

Table (4.7) reports on a series of experiments where the basic design was changed in three ways: (i) the x_t's were generated as NID (0,1) variates (and then kept fixed in repeated runs), (ii) $\beta = [2,10]'$ was used, and (iii) both modifications were applied at the same time. Also, we confine ourselves to the dynamic CUSUM test, since it has proven superior in the previous experiments.

Table 4.6: EMPIRICAL POWER OF THE TESTS (T = 120)

		static C -test			Dufour test			dynamic C -test		
					angle ψ					
	b	$0°$	$36°$	$90°$	$0°$	$36°$	$90°$	$0°$	$36°$	$90°$
d=0.3	4.8	20.7	14.3	3.2	6.9	6.1	3.4	10.5	7.7	5.3
	7.2	48.9	33.2	3.0	10.3	10.5	3.0	25.2	17.9	3.3
	9.6	78.3	52.4	2.1	8.8	15.6	3.7	32.8	28.0	5.2
	12.0	95.0	77.0	1.7	8.9	18.7	5.4	37.7	37.8	3.8
d=0.5	4.8	15.0	11.3	3.4	5.7	2.8	2.9	11.5	8.7	3.5
	7.2	38.9	22.2	2.7	3.1	4.6	4.1	19.5	13.4	2.4
	9.6	66.5	37.4	1.7	3.7	5.8	4.8	30.1	22.4	3.5
	12.0	87.7	59.9	1.3	1.9	6.3	4.2	33.5	32.6	1.4
d=0.7	4.8	6.7	4.7	3.5	2.1	2.4	4.1	6.2	4.8	3.1
	7.2	16.3	9.0	3.3	2.0	1.8	3.1	8.2	6.6	2.6
	9.6	28.5	14.5	1.9	1.0	1.3	4.3	11.5	9.6	2.3
	12.0	43.5	26.6	1.6	0.3	1.2	4.3	19.8	12.7	1.1

Table 4.7: THE DYNAMIC CUSUM TEST FOR ALTERNATIVE REGRESSORS AND
$\beta's$ (T = 120)

		x_t nid(0,1)			$\beta = [2,10]'$			both		
					angle Ψ					
	b	$0°$	$36°$	$90°$	$0°$	$36°$	$90°$	$0°$	$36°$	$90°$
d=0.3	4.8	16.5	13.0	2.9	12.0	10.4	2.0	20.7	11.5	3.6
	7.2	35.0	23.5	3.0	28.7	19.0	1.3	42.6	19.6	3.3
	9.6	49.4	33.0	2.0	52.0	34.4	1.6	63.8	37.6	3.7
	12.0	60.0	45.0	1.7	67.5	46.4	1.0	83.2	51.9	2.6
d=0.5	4.8	13.3	9.1	4.4	9.9	7.3	1.8	15.8	8.2	4.3
	7.2	24.2	16.2	3.5	24.2	14.2	1.3	28.9	17.6	2.1
	9.6	37.4	27.3	2.2	40.6	22.6	0.8	55.0	28.6	3.1
	12.0	50.4	36.0	2.4	55.9	34.3	0.4	70.8	45.4	2.0
d=0.7	4.8	6.9	6.2	2.8	4.7	5.4	2.1	7.2	5.5	3.9
	7.2	13.2	9.8	3.0	8.8	5.2	2.2	12.0	8.2	2.9
	9.6	19.7	14.1	4.4	15.6	6.4	1.2	20.4	13.4	3.1
	12.0	27.2	21.4	2.2	26.2	12.0	0.5	36.0	19.1	1.9

The basic pattern is similar to table (4.6), although the power is now generally higher for a given parameter combination. On the basis of this limited evidence, we tentatively conclude that there is not much reason in practice to use the Dufour procedure, and that one might stay with the straightforward CUSUM test also in dynamic models.

A modification of the CUSUM test

The CUSUM test requires that the cumulated sums of recursive residuals must be standardized by some estimate of the disturbance standard deviation σ. In the formula (4.9), we have followed standard practice by using

$$s = \left(\frac{1}{T\text{-}K} \Sigma_{t=1}^{T} \hat{u}_t^2 \right)^{1/2}, \tag{4.72}$$

where $\hat{u} = y\text{-}x\hat{\beta}$ is the vector of OLS regression residuals from the full sample. At first sight, this particular choice does not seem to matter much. The asymptotic null distribution of the test statistic (4.66) will be the same irrespective of the particular estimate for σ, as long as the latter is consistent.

Things are different under the alternatives, as pointed out by *Harvey* (1975). Depending on the type of structural change, the OLS residuals will often follow a very erratic pattern which will lead to a gross overestimation of σ^2 when using (4.72). Ceteris paribus, any overestimation of σ^2 will reduce the power of the test, since it is obvious from (4.9) that the cumulated sums of the recursive residuals are then less likely to cross the critical lines. This can even lead to a *decrease* in power when the intensity of the structural shift increases, as we have seen in some Monte Carlo experiments above.

Harvey (1975) therefore suggests to estimate σ by

$$\tilde{s} = \left(\frac{1}{T\text{-}K\text{-}1} \Sigma_{t=K+1}^{T} (\tilde{u}_t - \bar{\tilde{u}})^2 \right)^{1/2}, \tag{4.73}$$

where $\bar{\tilde{u}}$ is the arithmetic mean of the recursive residuals $\tilde{u}_t$. Under H_0, $\tilde{s}$ is consistent, but less efficient than s, since it does not incorporate the information that the population mean of the residuals is zero. In the presence of structural change, on the other hand, $\bar{\tilde{u}}$ might differ substantially from zero, and subtracting it from the recursive residuals will lead to a much smaller estimate for σ. Thus, to the extend that the overestimation of σ is mitigated, we will hopefully gain some power.

Next we examine this claim via some additonal Monte Carlo experiments. The design is as before, i.e.

$$y_t = 1 + (\text{-}1)^t + u_t ,$$

with NID(0,1) disturbances. The range of alternatives is widened somewhat, since we focus here on the effect of different types of structural shifts on the power of the test:

(i) Similar to the previous experiments, a single structural shift of the form (4.61), at times $T^* = 0.3T$, $T^* = 0.5T$ and $T^* = 0.7T$, respectively, (ii) two structural shifts, at time $T_1^* = 0.3T$ and $T_2^* = 0.7T$, respectively. Both shifts equal (4.61), i.e. they go in the same direction, (iii) as above, but the second shift now equals minus twice the first one, i.e. the shifts counteract each other, and (iv) the regression coefficients follow a random walk, i.e. $\beta_t = \beta_{t-1} + \eta_t \Delta\beta$, where the η_t are NID(0,1) (and kept fixed in repeated runs), and $\Delta\beta$ is as in (4.61).

Table (4.8) summarizes the results. As before, we vary the intensity b of the shift, and confine ourselves to a nominal significance level of $\alpha = 5\%$. We consider only the angle $\Psi = 36°$, since the effect of the angle Ψ on the power of the test has already been well established. 1000 trials were performed for any given parameter combination.

The table confirms the earlier pattern: the power of the test increases along with the intensity of the shifts, and is rather low when there is only one shift late in the sample period. The gain in power from an increase in the intensity of the shifts is rather small when the regression coefficients follow a random walk, i.e. when the structural changes are very erratic. Not surprisingly, when there are two shifts, with the second counteracting the first, the test is also rather weak, since then the recursive residuals will tend to have opposite signs in the respective regimes.

The most unambiguous result of the experiments is the increase in power obtained from the modification. The gain is greatest when the sample is small, and decreases in relative size as T increases. Therefore we conclude that there is certainly something to be gained, and nothing to be lost, by adopting Harvey's modification of the CUSUM test.

A modification of the CUSUM of squares test

Krämer et al. (1985) have shown that normality of the disturbances is not necessary in order to establish the null distribution of the CUSUM test. Thus it is natural to ask whether the same holds for the CUSUM of squares test. The answer is that the asymptotic null distribution of the test statistic

$$\tilde{S}^{(T)} = \max_{t=K+1,\ldots,T} \frac{\sum_{s=K+1}^{t} \tilde{u}_s^2}{\sum_{s=K+1}^{T} \tilde{u}_s^2} - \frac{t-K}{T-K} \tag{4.74}$$

is likewise robust to the presence of lagged dependent variables, but that the normality assumption is essential. This is a property that the CUSUM of squares test shares with e.g. the Breusch-Pagan test for heteroskedasticity (see section 3.b), whose asymptotic null distribution also depends on the normality of the regression disturbances. More generally, whenever a test is based on the *squares* of whichever type of residuals, the normality assumption becomes important.

Table 4.8: EMPIRICAL REJECTION RATES FOR VARIOUS STRUCTURAL SHIFTS

	b	single shift			two shifts		rand-walk
		$d=0.3$	$d=0.5$	$d=0.7$	$\Delta\beta_2=\Delta\beta_1$	$\Delta\beta_2=-2\Delta\beta_1$	
				a) $T = 30$			
	4.8	10.6	5.6	3.1	16.1	2.8	38.3
stand.	7.2	19.8	9.4	3.0	36.0	1.8	47.1
test	9.6	29.7	17.5	3.6	57.8	0.7	49.4
	12.0	48.0	25.0	4.5	74.0	0.4	49.8
	4.8	18.1	12.3	6.6	30.2	3.5	48.3
mod.	7.2	33.1	19.5	6.7	60.3	2.0	55.4
test	9.6	48.2	32.5	11.6	80.7	0.8	58.6
	12.0	67.5	44.6	16.2	91.8	0.4	58.5
				b) $T = 60$			
	4.8	12.6	8.6	4.9	27.1	7.2	67.6
stand.	7.2	31.4	16.9	7.7	57.1	8.1	74.3
test	9.6	47.2	32.1	10.5	85.0	8.2	73.9
	12.0	65.6	45.6	13.8	96.5	7.2	76.4
	4.8	16.0	11.4	6.3	34.1	7.8	70.4
mod.	7.2	36.7	21.3	10.7	66.8	8.5	77.1
test	9.6	55.8	39.4	15.3	89.7	8.5	76.7
	12.0	72.4	54.4	19.1	98.5	7.2	78.9
				c) $T = 120$			
	4.8	15.4	8.9	5.1	32.5	9.3	87.5
stand.	7.2	32.8	20.6	9.7	71.7	15.9	91.7
test	9.6	56.2	41.7	12.9	92.6	22.2	92.6
	12.0	78.6	61.1	25.4	99.6	30.6	91.5
	4.8	16.7	9.7	5.7	36.9	9.9	88.4
mod.	7.2	35.9	24.0	11.6	75.0	16.7	92.2
test	9.6	59.6	45.1	16.0	94.1	22.7	93.1
	12.0	81.1	66.9	29.0	99.7	31.3	92.3

The reason is that the kurtosis

$$k = E(u_t^4)/\sigma^4 \qquad (4.75)$$

enters the limiting distribution of the test statistics in some way or the other. The kurtosis equals $k = 3$ when the disturbances are normal, and only when this equality is preserved, will the asymptotic null distribution be invariant under changes of the distribution of the disturbances.

To see this for the CUSUM of squares test, let $Eu_t^4 = \mu$ and consider the statistic

$$\sqrt{T} \frac{\sigma^2}{\sqrt{\mu-\sigma^4}} \tilde{S}^{(T)}. \qquad (4.76)$$

This differs from the test statistic (4.74) of the CUSUM of squares test by the factor $\sqrt{T}\sigma^2/\sqrt{\mu-\sigma^4}$. Moreover, (4.76) has a well defined non-degenerate limiting distribution, with distribution function 1-f(c), where f is from (4.51), i.e.

$$P(\tilde{S} > x) = \Sigma_{j=1}^{\infty} 2 (-1)^{j-1} \exp(-2j^2 x^2). \qquad (4.77)$$

We show this by using the result from *Ploberger* and *Krämer* (1986b) that it does not matter asymptotically whether residuals or true disturbances are used in (4.74) and (4.76), provided only that the disturbances u_t are stationary martingale differences (where the conditioning σ-fields are generated by $\{x_{t-s}, y_{t-s-1}, u_{t-s-1} (s>0)\}$), with $E(u_t^2) = \sigma^2$ and $Eu_t^4 = \mu < \infty$. We can thus argue in terms of the simpler statistics

$$S^{(T)} = \max_{t=K+1,\dots,T} |S_t^{(T)}|,$$

where

$$S_t^{(T)} = \frac{\Sigma_{s=K+1}^{t} u_s^2}{\Sigma_{s=K+1}^{T} u_s^2} - \frac{t-K}{T-K}$$

$$= \frac{\sqrt{T}}{\Sigma_{s=K+1}^{T} u_s^2} \{\frac{1}{T} \Sigma_{s=K+1}^{t} (u_s^2 - \sigma^2) - \frac{1}{\sqrt{T}} \frac{t-K}{T-K} \Sigma_{s=K+1}^{T} (u_s^2 - \sigma^2)\}. \qquad (4.78)$$

The derivation of the limiting distribution of $\sqrt{T}S^{(T)}$ again requires some theory of weak convergence of random elements on the space D[0,1]. Similar to our discussion of the CUSUM test in dynamic models, let $\tau(z)$ be the largest integer less then or equal to $K + z(T-K)$, given a real number z ($0 \le z \le 1$), and consider the following stochastic process (random element) defined on the unit interval:

$$V^{(T)}(z) = \frac{1}{\sqrt{T}} \Sigma_{s=K+1}^{\tau(z)} (u_s^2 - \sigma^2). \qquad (4.79)$$

Then

$$\sqrt{T} S_t^{(T)} = \frac{T}{\Sigma_{s=K+1}^{T} u_s^2} \{V^{(T)}(z) - zV^{(T)}(1)\}. \qquad (4.80)$$

By assumption, $u_s^2 - \sigma^2$ is a martingale difference sequence with finite and constant second moment, so the random elements $V^{(T)}(z)$ tend in distribution to Brownian Motion with variance $(\mu-\sigma^4)z$ (see e.g. Billingsley, 1968, chapter 4). Consequently, the random elements

$$W^{(T)}(z) = V^{(T)}(z) - zV^{(T)}(1)$$

tend in distribution to a Brownian Bridge $W(z)$. The distribution theory of Brownian Bridges is well established and shows in particular that

$$\sup_{0 \leq z \leq 1} \frac{|W(z)|}{\sqrt{\mu - \sigma^4}}$$

has distribution function (4.77). Since

$$\sqrt{T}\, S^{(T)} = \frac{T}{\sum_{s=K+1}^{T} u_s^2} \sup_{0 \leq z \leq 1} |W^{(T)}(z)| \tag{4.81}$$

and

$$\frac{T}{\sum_{s=K+1}^{T} u_s^2} \xrightarrow{p} \frac{1}{\sigma^2},$$

the statistic

$$\sqrt{T}\, \frac{\sigma^2}{\sqrt{\mu - \sigma^4}}\, S^{(T)}$$

must also have the limiting distribution (4.77), which is the desired result.

The point of this proposition is that the limiting distribution of $\sqrt{T}\, S^{(T)}$, and thus of $\sqrt{T}\, \tilde{S}^{(T)}$, is unique only up to $\sigma^2/\sqrt{\mu - \sigma^4}$. This factor equals $1/\sqrt{2}$ for the normal distribution, but will differ for many other disturbance distributions. We therefore suggest to "studentize" the CUSUM of squares test by multiplying the test statistic $S^{(T)}$ with

$$\sqrt{T}\, \frac{\hat{\sigma}^2}{\sqrt{\hat{\mu} - \hat{\sigma}^4}},$$

where

$$\hat{\sigma}^2 = \sum_{t=K+1}^{T} \tilde{u}_t^2 /(T\text{-}K) \quad \text{and}$$

$$\hat{\mu} = \sum_{t=K+1}^{T} \tilde{u}_t^4 /(T\text{-}K)$$

Any other consistent estimates for μ and σ^2 would likewise produce a feasible version of the test statistic (4.76). Given that $\hat{\sigma}^2 \xrightarrow{p} \sigma^2$ and $\hat{\mu} \xrightarrow{p} \mu$, the limiting distribution of the statistic (4.76) is always given by (4.77), whether the disturbances are normal or not.

b) Functional Form

An implicit but basic assumption in the regression model (2.1) requires that the independent variables enter the relationship in a linear way. This is often anything but obvious.

When the true mathematical form of a relation is nonlinear but known, linearity can sometimes be achieved by appropriate transformation of the data. The standard example is the Cobb-Douglas production function, where linearity is obtained by taking logarithms on both sides of the equation. However, this is an exception, since the underlying theory only rarely provides any guidance as to the proper functional form, and the linear one is then chosen mostly for statistical and computational convenience. Needless to say that this is bound to produce all kinds of erraneous "results".

Ramsey's RESET

By far the most popular diagnostic for correctness of functional form is the RESET procedure (for "Regression Specification Error Test") proposed by *Ramsey* (1969) and earlier in *Anscombe* (1961). Its basic assumption is that under the alternative, the model can be written as or approximated by the regression

$$y = X\beta + Z\gamma + u \,. \tag{4.82}$$

This differs from the initial model (2.2) in that the regressor matrix is augmented by some $T \times p$ matrix Z.

Consider for illustration the bivariate model $y_t = \beta_1 + \beta_2 x_t + u_t$, where the true relationship is quadratic. Obviously, this means adding a column of x_t^2's to the initial regressor matrix X.

By expressing the alternative as in (4.82), it is clear that a wrong functional form is a special case of the more general problem of omitted variables. We will return to this topic in more detail in chapter 5 below, where the "omitted variable" theme will emerge as a red thread running through most of the econometric test methodology.

Given the alternative (4.82), RESET is simply the conventional F-test of H_0: $\gamma = 0$. In Ramsey's initial formulation, the test was based on a regression of the BLUS residuals from the restricted model on Z. *Ramsey* and *Schmidt* (1976) however show that this is equivalent to the F-test in (4.82), with test statistic

$$F = \tilde{\gamma}\,'\hat{V}^{-1}\tilde{\gamma}/p \,,$$

where $\tilde{\gamma}$ is the OLS estimate for γ and $\hat{V}$ is the OLS estimate for the covariance matrix of $\tilde{\gamma}$.

The crucial issue with RESET is the proper choice of the auxiliary variables for the Z-matrix. This matrix in turn depends on the true functional form under the alternative, which is usually unknown. However, it can often be well approximated by higher powers of the initial regressors as in the quadratic function above (see Ramsey, 1969). Alternatively, one might approximate the true functional form by adding higher moments of $\hat{y}_t = x_t'\hat{\beta}$ or moments of principle components of X (see *Thursby* and *Schmidt*, 1977). In the omitted variable context, this corresponds to situations where only proxies rather than true omitted variables are available.

This uncertainty does not affect the null distribution of the test statistic, which is always $F_{p,T-K-p}$, whether (4.82) is the true alternative or not. The only issue in the selection of the Z matrix concerns the power of the test when H_0 is wrong. Some choices of Z will then result in more powerful tests than other choices. Thursby and Schmidt (1977) conclude from Monte Carlo experiments that the test based on squared, cubed and fourth powers of the independent variables seems generally best.

The Rainbow test

A procedure somehow related to the RESET test is what Jessica Utts (1982) has called the "Rainbow Test" for lack of fit in the regression (she calls it rainbow test since it is supposed to cover a wide spectrum of possible problems in addition to nonlinearity). Similar to RESET, there is no need to specify a particular alternative.

The basic idea behind the rainbow test is that even when the true relationship is nonlinear, a good linear fit can still be obtained over subsets of the sample. The test therefore rejects the null hypothesis of linearity whenever the overall fit is markedly inferior to the fit over a properly selected subsample of the data.

Consider for instance the bivariate model depicted in figure 4.5, where the true relationship is quadratic but a straight line has been fit. If another straight line is fit only to the subset of the data points corresponding to the central region of the independent variable, then the linear fit is not as far from the true model as it is when fit over the entire range of observations.

Let $\tilde{u}$ be the OLS residual vector from the subset regression (T_1 observations), and let $\hat{u}$ always denote the overall OLS residual vector. The test statistic is then

$$F = \frac{(\hat{u}'\hat{u} - \tilde{u}'\tilde{u})/(T-T_1)}{\tilde{u}'\tilde{u}/(T_1-K)} \quad , \tag{4.83}$$

which is easily seen to have an F_{T-T_1, T_1-K} distribution under H_0.

The crucial issue with the rainbow test is the proper choice of the subsample. Again this concerns only the power of the test, and does not effect the null distribution of the test statistic. Utts (1982) recommends points close to $\bar{x}$, since an incorrect linear fit will in general not be as far off there as it is in the outer region. Closeness to $\bar{x}$ is measured by the magnitude of the corresponding diagonal element of the matrix $X(X'X)^{-1}X'$, i.e. the subset consists of data points with the smallest Mahalanobis distance from $\bar{x}$ (*Hoaglin* and *Welsch* 1978).

FIG.4.5: PARTIAL FIT VERSUS OVERALL FIT

The optimal size of the subset depends on the alternative. Utts recommends about one half of the data points in order to obtain some robustness to outliers.

The test statistic (4.83) is identical to the modified Chow test from (4.3). The expressions (4.83) and (4.3) differ only with respect to the criterion for selecting a subsample. The Chow test arranges the data points according to time and selects the first T^* ones; the rainbow test arranges the data according to their distance from $\bar{x}$ and selects the first T_1 ones. Otherwise the procedures are identical.

Convex or concave alternatives

The RESET and rainbow tests are often outperformed when something is known about the possible type of nonlinearity. A very simple procedure by *Harvey* and *Collier* (1977) addresses the case where the true functional form in some regressor is possibly convex (or concave) rather than linear. The basic idea is to rearrange the observations in ascending (or descending) order according to the variable which is to be tested for functional misspecification, and then to sum up the recursive residuals. When the true relationship is convex, recursive residuals will all tend to be positive, since they are simply the successive (standardized) forecasting errors. This is shown for the bivariate case in figure 4.6. Conversely, when the function is concave, the recursive residuals will tend to be negative, but in either case their sum will be large in absolute value.

FIG.4.6: RECURSIVE RESIDUALS WITH A CONVEX FUNCTION

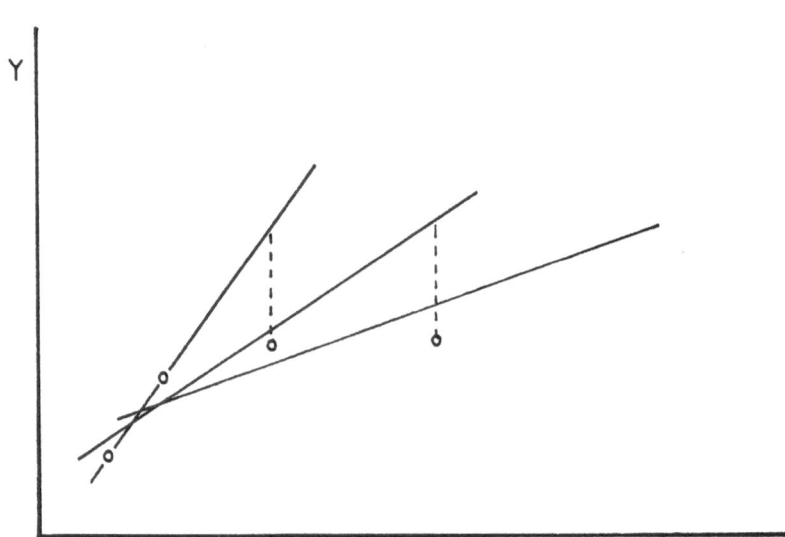

Let $\tilde{u}_t$ $(t = K+1,...,T)$ be the recursive residuals from (2.16), as obtained from the rearranged observations. The test statistic then is

$$\Psi = \frac{(\Sigma^T_{t=K+1} \hat{u}_t)/(T-K)^{1/2}}{[\Sigma^T_{t=K+1}(\tilde{u}_t-\bar{\tilde{u}})^2/(T-K-1)]^{1/2}} \quad , \qquad (4.84)$$

which follows a t-distribution with T-K-1 degrees of freedom under H_0. The null hypothesis of a linear functional form is rejected whenever $|\Psi|$ is too large.

The logic behind the Harvey-Collier Ψ-test is similar to the CUSUM test. Both procedures exploit particular patterns of recursive residuals which are likely to emerge under the respective types of misspecification. They differ in that the Ψ-test uses only the sum, and the CUSUM test also the partial sums, of the recursive residuals.

The reasoning that has led to the test statistic (4.84) does not carry over to situations where the functional form of more than one regressor is in doubt. Only when all of these enter in a convex (or concave) manner, and are positively correlated with each other, can one expect to obtain recursive residuals of the same sign (after rearranging the sample according to one of the suspect variables). Otherwise, the power of the Ψ-test will be rather low.

Harvey and Collier also consider the frequency of positive recursive residuals. This has a binomial distribution with parameter $p = 1/2$ and $n = T-K$ under H_0, and will be either large or small under the alternative.

Linear versus log-linear functional form

A second and quite unrelated type of prior information on the type of possible functional misspecification arises when some variables enter the model only after a Box-Cox transformation.

Godfrey and *Wickens* (1981) use this information for a straightforward Lagrange Multiplier test for the correctness of either a linear or a log-linear specification.

Given some scalar parameter λ, the Box-Cox transformation of x is

$$x(\lambda) = (x^{\lambda}-1) \, / \, \lambda \; . \tag{4.85}$$

The expression (4.85) is not defined for $\lambda = 0$, but it is easily seen (provided $x > 0$) that

$$\lim_{\lambda \to 0} x(\lambda) = \ln(x) \; , \tag{4.86}$$

so we define $x(0) := \ln(x)$.

Consider now instead of (2.1) the more general regression model

$$y_t(\lambda) = \Sigma_{i=1}^{K_1} \; \beta_i \, x_{ti}(\lambda) + \Sigma_{i=K_1+1}^{K} \beta_i z_{ti} + u_t \quad (t=1,...,T) \; , \tag{4.87}$$

where the z's are seasonal or other dummies or more general any variables which are known to enter the equation without any prior transformation. We assume here that these always include a vector of ones to account for a constant term. Then we obtain a linear model by setting $\lambda = 1$, since $x_{ti}(1) = x_{ti}-1$ and the -1 is absorbed by the constant term. Similarly, we obtain a log-linear model by setting $\lambda = 0$ (provided all y_t's and x_{ti}'s are positive). The acceptability of the linear and the log-linear models can thus be investigated by testing the null hypotheses $H_0 : \lambda = 1$ and $\bar{H}_0 : \lambda = 0$, respectively.

Given that the u_t's are NID($0,\sigma^2$), the log likelihood for the t'th observation in (4.87) is

$$l_t(\theta) = (\ln 2\pi)/2 - (\ln \sigma^2)/2 + (\lambda-1) \ln y_t$$
$$- [y_t(\lambda) - \Sigma_{i=1}^{K_1} \beta_i \, x_{ti}(\lambda) - \Sigma_{i=K_1+1}^{K} \beta_i z_{ti}]^2 \, / \, 2\sigma^2 \; , \tag{4.88}$$

where $\theta = [\beta ',\sigma^2, \lambda]'$, and the term $(\lambda-1)\ln y_t$ is the logarithm of the derivative of the function that transforms y_t into $y_t(\lambda)$.

Godfrey and Wickens (1981) use (2.45) to approximate the information matrix that we need for the LM test, so we next need $\partial l_t / \partial \theta$, evaluated at the respective restricted maximum likelihood estimates. Let us first consider the test for H_0: $\lambda = 1$. Here the restricted ML estimate for θ is $\bar{\theta} = [\hat{\beta}, \hat{\sigma}^2, 1]$, where $\hat{\beta}$ and $\hat{\sigma}^2$ are obtained from OLS.

Moreover,

$$\partial l_t(\bar{\theta})/\partial\beta_i = (x_{ti}-1) \, \hat{u}_t/\hat{\sigma}^2 \quad (i=1,...,K_1) \; , \tag{4.89}$$

$$\partial l_t(\bar{\theta})/\partial\beta_i = z_{ti} \, \hat{u}_t/\hat{\sigma}^2 \quad (i=K_1+1,...,K) \; , \tag{4.90}$$

$$\partial l_t(\bar{\theta})/ \partial\sigma^2 = (\hat{u}_t^2 - \hat{\sigma}^2)/2\hat{\sigma}^4 \; , \text{ and} \tag{4.91}$$

$$\partial l_t(\bar{\theta})/\partial\lambda = \ln y_t - [(y_t \ln y_t - y_t + 1 - \Sigma_{i=1}^{K_1} \hat{\beta}_i \, (x_{ti} \ln x_{ti} - x_{ti} + 1)) \, \hat{u}_t/\hat{\sigma}^2] \; . \tag{4.92}$$

These expressions may look complicated, but are straightforwardly derived from (4.88). They immediately produce the $T \times (K+2)$ matrix $W(\bar{\theta})$ with typical element $w_{tk} = \partial l_t(\bar{\theta})/\partial \theta_k$ that we need to form the LM test statistic (2.46). In the present context, this has an asymptotic $\chi^2_{(1)}$ distribution under H_0, since there is only one restriction under test.

When testing $\tilde{H}_0: \lambda = 0$, we have $\bar{\theta} = [\bar{\beta}', \bar{\sigma}^2, 0]'$, where now $\bar{\beta}$ and $\bar{\sigma}^2$ are the OLS estimates from the logarithmic model, and where (4.89) - (4.92) are replaced by

$$\partial l_t(\bar{\theta})/\partial \beta_i = (\ln x_{ti}) \, \hat{u}_t/\hat{\sigma}^2 \quad (i = 1,...,K_1) , \tag{4.93}$$

$$\partial l_t(\bar{\theta})/\partial \beta_i = z_{ti} \, \hat{u}_t/\hat{\sigma}^2 \quad (i = K_1 + 1,...,K) , \tag{4.94}$$

$$\partial l_t(\bar{\theta})/\partial \sigma^2 = (\hat{u}_t^2 - \hat{\sigma}^2)/2\hat{\sigma}^4 , \text{ and} \tag{4.95}$$

$$\partial l_t(\bar{\theta})/\partial \lambda = \ln y_t - [((\ln y_t)^2/2 - \Sigma_{i=1}^{K_1} \, \hat{\beta}_i (\ln x_{ti})^2/2) \, \hat{u}_t/\hat{\sigma}^2] . \tag{4.96}$$

Again the asymptotic null distribution of the resulting test statistic is $\chi^2_{(1)}$ since only one restriction is being tested.

Outliers

A problem which might also be viewed as a misspecification of the functional form are outliers in the regression. We confine ourselves to the case where just one observation, the t'th say, is suspected as being an outlier. Moreover, we focus on the mean-shift outlier model (see e.g. *Cook* and *Weisberg*, 1982, sec.2.2.2) which can be written as

$$y = X\beta + d_t\gamma + u \tag{4.97}$$

where d_t is a T vector with t'th element equal to one, and all other elements equal to zero.

Obviously, there is no outlier when $\gamma = 0$. We therefore reject H_0 whenever $|\tau_t| > t_{\alpha/2, T-K-1}$, where τ_t is the t-value for γ and where $t_{\alpha/2, T-K-1}$ is the critical $\alpha/2$ value for the central t-distribution with T-K-1 degrees of freedom.

A major problem arises when t is unknown, which will usually be the case. Common practice then is to base the decision on $\max_{1 \leq t \leq T} |\tau_t|$. In our own empirical work below, we reject H_0 whenever

$$\max_{1 \leq t \leq T} |\tau_t| > t_{\alpha/2T, T-K-1} \ . \tag{4.98}$$

This ensures that the probability of a type I error is bounded by α, since

$$P(\max|\tau_t| > t_{\alpha/2T, T-K-1})$$

$$= P(\bigcup_{t=1,...,T} \{|\tau_t| > t_{\alpha/2T, T-K-1} \}) \tag{4.99}$$

$$\leq \Sigma_{t=1}^{T} P(|\tau_t| > t_{\alpha/2T, T-K-1})$$

$$= \Sigma_{t=1}^{T} \alpha/T$$

$$= \alpha \ .$$

c) General Misspecification

This section is concerned with tests that do not require a specific alternative. Such procedures are often called "pure significance tests", since a significant test result is simply taken as evidence that something is wrong with the model, without exactly specifying what this might be.

However, the distinction that is thus drawn between pure significance tests and tests which are based on a specific alternative is rather artificial and does not derive from any intrinsic properties of the tests themselves. Whether or not to apply the label "pure significance test" rather depends on the a priori information available in a particular application.

For instance, if AR(1) disturbances are the only possible violation of the ideal assumptions in our linear model, than a significant Durbin-Watson statistic unambiguously points to autocorrelation among the disturbances. However, if we widen the range of possible violations of the model assumptions, we do no longer necessarily regard a significant test result as implying that the disturbances are serially correlated, since the DW statistic is well known to be sensitive to various other departures from the null hypothesis. Whether or not the Durbin-Watson test should be viewed as a pure significance test thus depends on the range of possible alternatives, that is on our a priori information or what some call the "maintained hypothesis", and is not an intrinsic property of the test.

What distinguishes the procedures from this section from others is rather the lack of a well defined alternative in the construction stage of the test statistic. However this is a matter of degree, and procedures like the CUSUM test, the Harvey-Collier Ψ-test or the Rainbow test could with some justice also appear under the present heading.

Hausman-type specification tests

Consider the following simple but fundamental idea: If a model is correctly specified, estimates by any two consistent methods should be close to each other. If not, there is reason to believe that something is wrong. This idea had been applied by *Durbin* (1954) and *Wu* (1973) to the special case of testing for the independence between stochastic regressors and disturbance terms in the standard linear model, but its full potential was brought to light only recently by *Hausman* (1978). There has been a growing number of applications ever since.

To formalize this idea, let $\hat{\beta}$ and $\tilde{\beta}$ be consistent estimates for the coefficient vector $\beta(K \times 1)$ in some econometric model (not necessarily the linear regression model), and assume that

$$\sqrt{T} (\beta - \tilde{\beta}) \xrightarrow{d} N(0,V) \tag{4.100}$$

as $T \to \infty$, where V is the covariance matrix of the asymptotic distribution of $\sqrt{T}(\hat{\beta} - \tilde{\beta})$. Assume for the moment that V is nonsingular, and that a consistent estimate $\hat{V}$ for V is available (below we will see that this is a major stumbling block in empirical applications of the Hausman test). Let $\hat{q} = \hat{\beta} - \tilde{\beta}$. A straightforward test then rejects the model whenever the statistic

$$m = T\hat{q}' (\hat{V}^{-1}) \hat{q} \tag{4.101}$$

is too large, where the appropriate rejection region is obtained from the easy fact that m has an asymptotic $\chi^2_{(K)}$ distribution under H_0.

This general approach can be applied in many different situations. The resulting test will be consistent against any alternative which produces different probability limits for $\hat{\beta}$ and $\tilde{\beta}$, since the test statistic (4.101) will then tend to infinity in probability as $T \to \infty$. The problem in a given application is to find two such estimates, and to consistently estimate the covariance matrix of the asymptotic distribution of the difference.

Given a particular alternative, Hausman (1978) suggests choosing the estimates such that $\hat{\beta}$ is asymptotically efficient under H_0, but inconsistent under H_1, and such that $\tilde{\beta}$ is consistent under both H_0 and H_1. Moreover, $\hat{\beta}$ and $\tilde{\beta}$ should be asymptotically jointly normal under H_0.

Then plim $\hat{\beta} \neq$ plim$\tilde{\beta}$ under H_1 in view of the inconsistency of $\hat{\beta}$, and the covariance matrix of the asymptotic null distribution of the difference will simply be the difference of the covariance matrices of the respective asymptotic distributions of the estimates.

This latter result follows from the asymptotic efficiency of $\hat{\beta}$. It can be shown (see Hausman, 1978, Lemma 2.1) that the joint limiting distribution of $\sqrt{T}(\hat{\beta}-\beta)$ and $\sqrt{T}\hat{q}$ must then have a zero covariance matrix (since $\hat{\beta}$ could otherwise be approved upon), i.e. that $V(\tilde{\beta}) = V(\hat{q}) + V(\hat{\beta})$ and thus

$$V(\hat{q}) = V(\tilde{\beta}) - V(\hat{\beta}), \tag{4.102}$$

where $V(\tilde{\beta})$ is the covariance matrix of the asymptotic distribution of $\sqrt{T}(\tilde{\beta}-\beta)$, and $V(\hat{\beta})$ is the covariance matrix of the asymptotic distribution of $\sqrt{T}(\hat{\beta}-\beta)$.

The relationship (4.102) forms the heart of all Hausman tests and is often called the Hausman Lemma. It is the major reason for the popularity of the procedure, since consistent estimates for $V(\tilde{\beta})$ and $V(\hat{\beta})$ are in general easily obtained and can then via (4.102) be used to consistently estimate $V(\hat{q})$.

More often than not, $\hat{\beta}$ equals OLS. Moreover, the comparison estimator $\tilde{\beta}$ is often chosen without regard to a particular alternative, which is why Hausman tests are grouped among tests for general misspecification. The alternative is then implicitly defined by all models which lead to different probability limits for $\tilde{\beta}$ and $\hat{\beta}$.

Errors in variables

Let us slightly generalize the standard linear regression model by allowing for stochastic elements in the regressor matrix X. We still assume that all elements of X are independent of all elements in the disturbance vector u (not just the contemporaneous ones). This ensures that standard procedures such as the F- and t-tests are still valid. Let $\hat{\beta}$ be OLS, and assume

$$\text{plim}_{T-\infty} (1/T) X'X = Q \quad \text{(nonsingular)} .$$

(4.103)

Then

$$\sqrt{T} (\hat{\beta} - \beta) \xrightarrow{d} N(0,\sigma^2 Q^{-1})$$

(4.104)

and OLS is asymptotically efficient.

Now decompose the regression equation (2.2) as

$$y = X^{(1)} \beta^{(1)} + X^{(2)} \beta^{(2)} + u$$

(4.105)

where $X^{(1)}$ $(T \times K_1)$ comprises the stochastic and $X^{(2)}$ $(T \times (K-K_1))$ comprises the nonstochastic columns of X. Under the alternative, $X^{(1)}$ is measured with error, i.e. the true mechanism that generates y is

$$y = \tilde{X}^{(1)} \beta^{(1)} + X^{(2)} \beta^{(2)} + \epsilon ,$$

(4.106)

where $X^{(1)} = \tilde{X}^{(1)} + W$ with some stochastic error matrix W. We assume that the rows w_t of the error matrix W are serially independent with expectation zero and covariance matrix $\Omega(K_1 \times K_1)$, and that $E(w_w \epsilon_t) = 0$. The null hypothesis can therefore be expressed as $H_0 : \Omega = 0$.

Substituting for $X^{(1)}$ in (4.105) implies that $u_t = \epsilon_t - w_t'\beta^{(1)}$, i.e. that

$$\text{cov}(X_t^{(1)}, u_t) = E(w_t u_t) = -\beta^{(1)}\Omega ,$$

(4.107)

which will in general be different from zero. Measurement errors in the regressors thus imply correlation between regressors and disturbances in the model (4.105), which in turn leads to the inconsistency of OLS.

To see this, consider $\hat{\beta} - \beta = (X'X)^{-1} X'u$, which implies that

$$\text{plim}_{T-\infty} \hat{\beta} - \beta = \{\text{plim} (X'X/T)\}^{-1} \text{plim} X'u/T$$

$$= Q^{-1} \begin{bmatrix} -\Omega\beta^{(1)} \\ 0 \end{bmatrix}.$$

(4.108)

This shows that the inconsistency of OLS will in general increase with increasing length of $\beta^{(1)}$, and that OLS will be consistent for $\beta^{(1)} = 0$.

Our Hausman test for errors in variables is now based on comparing $\hat{\beta}$ to some Instrumental Variables (IV) estimate $\tilde{\beta}$, defined as

$$\tilde{\beta} = (H'X)^{-1} H'y , \tag{4.109}$$

where H is $T \times K$ matrix of instruments.

There are many different IV estimators, depending on the choice of H. Regardless of H, $\tilde{\beta}$ is consistent and asymptotically normal, whether there are errors in variables or not, whenever

plim $_{T \to \infty}$ H'X/T is nonsingular,

plim $_{T \to \infty}$ H'H/T is nonsingular, and $\tag{4.110}$

plim $_{T \to \infty}$ H'u/T = 0

both under H_0 and H_1. The covariance matrix of the limiting distribution of $\sqrt{T}$ $(\tilde{\beta}-\beta)$ is then

$$V(\tilde{\beta}) = \sigma_u^2 (\text{plim } H'X/T)^{-1} (\text{plim } H'H/T) (\text{plim } X'H/T)^{-1} . \tag{4.111}$$

Below we assume (4.110). In addition we use $H=[\hat{X}^{(1)}, X^{(2)}]$, where $\hat{X}^{(1)} = Z(Z'Z)^{-1}Z'X^{(1)}$ are fitted values from a regression of $X^{(1)}$ on some auxiliary matrix Z (which in our case is made up of $X^{(2)}$ and lagged values of the variables in $X^{(1)}$).

Sometimes the columns of Z are also called instruments.

When H_0 is true, i.e. when there are no measurement errors in the independent variables, the asymptotic covariance matrix for OLS is $V(\hat{\beta})=\sigma^2(\text{plim } X'X/T)^{-1}$, and the asymptotic covariance matrix for IV is given by (4.111). From (4.102), we then have $V(\hat{q}) = V(\tilde{\beta})-V(\hat{\beta})$, which under H_0 is consistently estimated by

$$\hat{V}(\hat{q}) = \tilde{\sigma}^2 (H'X/T)^{-1}H'H/T (X'H/T)^{-1} - \hat{\sigma}^2(X'X/T)^{-1} \tag{4.112}$$

where $\tilde{\sigma}^2$ and $\hat{\sigma}^2$ are any two estimators for σ_u^2 that are consistent under H_0.

It is here that the trouble with the Hausman test begins. The reason is that more often than not, $V(\hat{q})$ will be singular, so that the test statistic cannot be computed as in (4.101).

Computing the Hausman test

The singularity of $V(\hat{q})$ is in the present context easily seen from

$$H'X = \begin{bmatrix} \hat{X}^{(1)\prime}X^{(1)} & \hat{X}^{(1)\prime}X^{(2)} \\ X^{(2)\prime}X^{(1)} & X^{(2)\prime}X^{(2)} \end{bmatrix}$$

$$= \begin{bmatrix} \hat{X}^{(1)\prime}\hat{X}^{(1)} & \hat{X}^{(1)\prime}X^{(2)} \\ X^{(2)\prime}\hat{X}^{(1)} & X^{(2)\prime}X^{(2)} \end{bmatrix}$$

$$= H'H$$

$$= X'H,\qquad(4.113)$$

which implies $(H'X)^{-1}H'H(X'H)^{-1} = (H'H)^{-1}$ and

$$V(\hat{q}) = \text{plim } \sigma^2 T((H'H)^{-1} -(X'X)^{-1}).\qquad(4.114)$$

However,

$$X'X - H'H = \begin{bmatrix} X^{(1)'}X^{(1)}-\hat{X}^{(1)'}\hat{X}^{(1)} & 0 \\ 0 & 0 \end{bmatrix}\qquad(4.115)$$

has at most rank K_1, so the same applies to $H'H^{-1}-X'X^{-1}$. This follows from

$$H'H^{-1}- X'X^{-1} = X'X^{-1}(X'X - H'H) H'H^{-1}\qquad(4.116)$$

and the well known fact that the rank of a product of matrices cannot exeed the rank of any of the factors.

The singularity of $V(\hat{q})$ can also be seen by expressing $V(\hat{q})$ in more detail as

$$V(\hat{q}) = \text{plim } \sigma^2 T((H'H)^{-1}-(X'X)^{-1})$$

$$= \text{plim } \sigma^2 T\begin{bmatrix} \bar{\Sigma} - \hat{\Sigma} & (\hat{\Sigma} - \bar{\Sigma})A \\ A'(\hat{\Sigma}- \bar{\Sigma}) & A'(\bar{\Sigma} - \hat{\Sigma})A \end{bmatrix}\qquad(4.117)$$

where

$$A = X^{(1)'}X^{(2)} (X^{(2)'}X^{(2)})^{-1},\qquad(4.118)$$

$$\bar{\Sigma} = [(\hat{X}^{(1)'}\hat{X}^{(1)} -X^{(1)'}X^{(2)} (X^{(2)'}X^{(2)})^{-1} X^{(2)'}X^{(1)}]^{-1}$$

and

$$\hat{\Sigma} = [X^{(1)'}X^{(1)} -X^{(1)'}X^{(2)}(X^{(2)'}X^{(2)})^{-1} X^{(2)'}X^{(1)}]^{-1}\qquad(4.119)$$

The expression (4.117) for $V(\hat{q})$ follows from a straightforward application of the well known inversion formula for partitioned matrices (see e.g. Theil, 1971, p.18), which implies

$$(H'H)^{-1} = \begin{bmatrix} \bar{\Sigma} & - \bar{\Sigma}A \\ -A'\bar{\Sigma} & (X^{(2)'}X^{(2)})^{-1} + A'\bar{\Sigma}A \end{bmatrix}\qquad(4.120)$$

A similar formula holds for $(X'X)^{-1}$, with $\bar{\Sigma}$ replaced by $\hat{\Sigma}$.

From (4.117) it is clear that the rank of $V(\hat{q})$ will at most equal the rank of plim $T(\bar{\Sigma}-\hat{\Sigma})$, which in turn will always have full rank under the present assumptions. Defining $\Sigma = T(\bar{\Sigma}-\hat{\Sigma})$, it is in addition obvious from (4.117) that $\hat{\sigma}^2\Sigma$ is a consistent estimate for the covariance matrix of the limiting distribution of $\sqrt{T}(\bar{\beta}^{(1)}-\hat{\beta}^{(1)})$.

Using (4.116), it can more generally be shown that $V(\hat{q})$ will be nonsingular if and only if the intersection of the spaces spanned by the columns of X and Z has dimension zero, which will almost never hold in practice due to the existence of common columns in X and Z. The singularity of $V(\hat{q})$ is thus not peculiar to the present example but is rather a common feature of most IV-based Hausman tests. *Hausman* and *Taylor* (1981) and *Holly* (1982) have acknowledged this and have suggested to use a generalized inverse $\hat{V}(\hat{q})$ rather than $\hat{V}(\hat{q})^{-1}$ in the test statistic (4.101). If the g-inverse is chosen such that

$$\hat{V}(\hat{q})^{-} \overset{p}{\to} V(\hat{q})^{-}, \tag{4.121}$$

where $V(\hat{q})^{-}$ has rank $K_1 \leq K$, the modified test statistic will then still have an asymptotic χ^2- distribution under H_0, though with K_1 rather than K degrees of freedom.

In the present context, a convenient g-inverse of $\hat{V}(\hat{q})$ is

$$\hat{V}(\hat{q})^{-} = \hat{\sigma}^{-2} \begin{bmatrix} \Sigma^{-1} & 0 \\ 0 & 0 \end{bmatrix}. \tag{4.122}$$

The general test statistic (4.83) then reduces to

$$m = T(\tilde{\beta}^{(1)} - \hat{\beta}^{(1)})'(\hat{\sigma}^2 \Sigma)^{-1} (\tilde{\beta}^{(1)} - \hat{\beta}^{(1)}), \tag{4.123}$$

i.e. to a comparison of the coefficient estimates of the contaminated regressors. In this form the Hausman test has been known for some time in the simultaneous equation context.

A major advantage of the expression (4.122) is that the explicit computation of a generalized inverse can be avoided. Since (4.121) holds as well, the statistic (4.122) has an asymptotic $\chi^2_{(K_1)}$ -distribution under H_0.

The problems that arise from the possible singularity of $V(\hat{q})$ are thus solved in theory, but unfortunately not in practice. The reason is the crucial condition (4.121), which is often hard to verify.

To avoid all unnecessary complications with the nonuniqueness of generalized inverses, let us argue in terms of the Moore-Penrose inverse $V(\hat{q})^+$, which is always unique. (Note however that (4.122) is not the Moore-Penrose inverse, but rather a reflexive g-inverse, in the notation of *Rao* and *Mitra*, 1971). The problem is that g-inverses are not continuous, i.e. that from $\hat{V}(\hat{q}) \overset{p}{\to} V(\hat{q})$ we cannot infer

$$\hat{V}(\hat{q})^+ \overset{p}{\to} V(\hat{q})^+. \tag{4.124}$$

A sufficient condition for (4.124) is

$$\text{rank}(\hat{V}(\hat{q})) = \text{rank}(V(\hat{q})) \tag{4.125}$$

with probability one, for all $T > T_0$. This obviously holds in the present context in view of (4.122), but will not be satisfied in general.

The most obvious case where (4.125) fails is when $\hat{V}(\hat{q})$ has full rank irrespective of sample size, despite the singularity of its probability limit $V(\hat{q})$. This happens for instance whenever different estimates for σ^2 are used in (4.112).

In practice, one often uses $\tilde{\sigma}^2 = \tilde{u}'\tilde{u}/(T-K)$ and $\hat{\sigma}^2 = \hat{u}'\hat{u}/(T-K)$, where $\tilde{u} = y - X\tilde{\beta}$ and $\hat{u} = y - X\hat{\beta}$. Then plim $\tilde{\sigma}^2 = $ plim $\hat{\sigma}^2 = \sigma^2$ under H_0, so $\hat{V}(\hat{q}) \overset{p}{\to} V(\hat{q})$, but there is no g-inverse $V(\hat{q})^{-}$ such that $\hat{V}(\hat{q})^{-1} \overset{p}{\to} V(\hat{q})^{-}$. In fact, it is easily verified that $\hat{V}(\hat{q})^{-1} \overset{p}{\to} \infty$ (at least for some components), so $\hat{V}(\hat{q})^{-1}$ does not converge to anything.

The following example shows that this can have serious implications for the limiting distribution of a related quadratic form: Let $K = 2$, $\hat{q} \sim N(0, V(\hat{q}))$ for all T, where $V(\hat{q})$ has rank 1, with a 1 in the upper left corner and zeros elsewhere. Let

$$\hat{V}(\hat{q}) = \begin{bmatrix} 1 & \dfrac{1}{\sqrt{T}} - \dfrac{1}{T^2} \\ \dfrac{1}{\sqrt{T}} - \dfrac{1}{T^2} & \dfrac{1}{T} \end{bmatrix} .$$

Then obviously $\hat{V}(\hat{q}) \xrightarrow{p} V(\hat{q})$, but $\hat{V}(\hat{q})^{-1} \xrightarrow{p} V(\hat{q})^{-1}$ and $\hat{q}'[\hat{V}(\hat{q})]^{-1}\hat{q}\xrightarrow{p}\infty$. In particular, the quadratic form does not have a limiting χ^2 distribution.

This example signals trouble for the Hausman test when (4.124) fails. Fortunately it is rather untypical.

The reason why the quadratic form in the above example tends to infinity is that $\hat{V}(\hat{q})^{-1}$ "explodes" as $T \to \infty$. This is the same with Hausman tests, but the dependency among the elements of $\hat{q}$ is often such that the limiting distribution of the test statistic remains intact.

So let $\tilde{\sigma}^2 \neq \hat{\sigma}^2$ with probability one. Then

$$\hat{V}(\hat{q}) = \tilde{\sigma}^2 (H'H/T)^{-1} - \hat{\sigma}^2 (X'X/T)^{-1}$$

$$= T \begin{bmatrix} \Omega & -\Omega A \\ -A'\Omega & C + A'\Omega A \end{bmatrix} , \tag{4.126}$$

where A is from (4.118), $\Omega = \tilde{\sigma}^2\tilde{\Sigma} - \hat{\sigma}^2\hat{\Sigma}$, and $C = (\tilde{\sigma}^2 - \hat{\sigma}^2)(X^{(2)}{}'X^{(2)})^{-1}$. Direct multiplication shows that

$$\hat{V}(\hat{q})^{-1} = T^{-1} \begin{bmatrix} \Omega^{-1} + AC^{-1}A' & AC^{-1} \\ C^{-1}A' & C^{-1} \end{bmatrix} . \tag{4.127}$$

Moreover,

$$\hat{q} = \tilde{\beta} - \hat{\beta} = [(H'H)^{-1}H' - (X'X)^{-1}X']y ,$$

where in view of (4.120)

$$(H'H)^{-1}H' = \begin{bmatrix} \tilde{\Sigma}\tilde{X}^{(1)}{}' - \tilde{\Sigma}AX^{(2)}{}' \\ -A'\tilde{\Sigma}\tilde{X}^{(1)}{}' + (X^{(2)}{}'X^{(2)})^{-1}X^{(2)}{}' + A'\tilde{\Sigma}AX^{(2)}{}' \end{bmatrix} \tag{4.128}$$

and

$$(X'X)^{-1}X' = \begin{bmatrix} \hat{\Sigma}X^{(1)}{}' - \hat{\Sigma}AX^{(2)}{}' \\ -A'\hat{\Sigma}X^{(1)}{}' + (X^{(2)}{}'X^{(2)})^{-1}X^{(2)}{}' + A'\hat{\Sigma}AX^{(2)}{}' \end{bmatrix} . \tag{4.129}$$

Thus,

$$(H'H)^{-1}H' - (X'X)^{-1}X' = \begin{bmatrix} W \\ -A'W \end{bmatrix} ,$$

where $W = \tilde{\Sigma}\tilde{X}^{(1)}{}' - \tilde{\Sigma}AX^{(2)}{}' - \hat{\Sigma}X^{(1)}{}' + \hat{\Sigma}AX^{(2)}{}'$ and

$$\hat{q} = \begin{bmatrix} \hat{q}^{(1)} \\ -A'\hat{q}^{(1)} \end{bmatrix} , \tag{4.130}$$

i.e. there is an exact linear relationship between the first K_1 elements of $\hat{q}$ and the rest.

Combining (4.127) and (4.130) shows that the general formula (4.101) for the test statistic can be reexpressed as

$$\tilde{m} = T\hat{q}' [\hat{V}(\hat{q})]^{-1}\hat{q}$$
$$= \hat{q}^{(1)}{}' \Omega^{-1}\hat{q}^{(1)}$$
$$= T(\tilde{\beta}^{(1)} - \hat{\beta}^{(1)})' (T\Omega)^{-1} (\tilde{\beta}^{(1)} - \hat{\beta}^{(1)}), \tag{4.131}$$

i.e. the test statistic can be written as a quadratic form in the first K_1 elements of $\hat{q}$. Moreover, when both $\tilde{\sigma}^2$ and $\hat{\sigma}^2$ are consistent,

$$(\tilde{\sigma}^2\Sigma)^{-1} - (T\Omega)^{-1} \xrightarrow{p} 0 , \tag{4.132}$$

so that the statistic (4.131) has under H_0 the same asymptotic $\chi^2_{(K_1)}$ distribution as (4.123).

This is a very reassuring result, since it shows that the discontinuity of the g-inverse does not matter in the present context. The only mistake that is often made in practice concerns the degrees of freedom of the test statistic. This is often erraneously taken to be K (see e.g. the papers by *Frenkel*, 1980a, 1980b, 1981), when in reality it is only K_1. This leads to rejection regions with a smaller asymptotic rejection probability than the nominal α level of the test.

Table 4.9 gives the true rejection probabilities for various values of α, K and K_1 and the critical χ^2 values corresponding to K and α . The figures show that depending on K and K_1, the overstatement of the true size of the test can often be dramatic, so the correct choice of the degrees of freedom of the test statistic is certainly not an unimportant matter.

Exogeneity

Measurement errors in the regressors lead to correlation among regressors and disturbances and thus to the inconsistency of OLS. The same happens when the equation under test is part of a simultaneous equation system and when some of the right-hand-side variables are endogenous in the larger system framework. The structure of the problem is the same and so the same test applies.

Let us briefly return to the simultaneous equation notation from section 3.a . Write the equation of interest (which we take without loss of generality to be the first) as

$$y_1 = Y_2 \beta + Z_1 \gamma + u_1 , \tag{4.133}$$

where as before y_1 and Y_2 are, respectively, $T \times 1$ and $T \times g_1$ matrices of observations on the $g_1 + 1$ endogenous variables in the equation. Z_1 ($T \times K_1$) contains the included (nonstochastic) exogenous variables, and Z ($T \times K$) contains all exogenous variables of the system.

Since u_1 can now be written as in (3.55), we again have correlation between some regressors (those in Y_2) and the disturbances of the equation, and OLS will be inconsistent under the usual assumptions on the large sample behaviour of $Z'Z$ (not however when there is a trend in the data; see *Krämer*, 1984a, 1985a).

Table 4.9: ACTUAL AND NOMINAL REJECTION PROBABILITIES UNDER H_0

df (actual)	degrees of freedom (nominal)					
	2	3	4	5	6	7
	a) $\alpha = 1\%$					
	9.21	11.36	13.28	15.09	16.81	18.48
1	.24	.08	.03	.01	.00	.00
2	1.00	.34	.13	.05	.02	.01
3		1.00	.41	.17	.08	.04
4			1.00	.45	.21	.10
	b) $\alpha = 5\%$					
	5.99	7.82	9.49	11.07	12.59	14.07
1	1.44	.52	.21	.09	.04	.02
2	5.00	2.01	.87	.39	.18	.09
3		5.00	2.35	1.13	.56	.28
4			5.00	2.58	1.35	.71
	c) $\alpha = 10\%$					
	4.61	6.25	7.78	9.24	10.65	12.02
1	3.19	1.24	.53	.24	.11	.05
2	10.00	4.39	2.05	.99	.49	.25
3		10.00	5.08	2.63	1.38	.73
4			10.00	5.55	3.09	1.72

The equation (4.133) is given in the form that is valid under the alternative. The null hypothesis we wish to test is that there is no correlation between Y_2 and u_1, i.e. that the variables in Y_2 are exogenous. Then OLS will be consistent and asymptotically efficient.

We proceed along the same lines as before. Let $\hat{\beta}$ be the OLS estimate for β (which is here only a subvector of the regression coefficients in the equation). This is now compared to the 2SLS-estimate β^* from (3.56). It is well known that 2SLS can be viewed as an IV estimator, where the H-matrix is $[\hat{Y}_2, Z_1]$ and $\hat{Y}_2 = Z(Z'Z)^{-1}Z'Y_2$ is obtained by regressing all columns of Y_2 on Z (in more technical jargon: $\hat{Y}_2$ is the orthogonal projection of Y_2 on the column space of Z).

In order for H to have full column rank, we require a sufficient number of linearly independent columns in Z (at least $g_1 + K_1$).

Now the formalism from above can be used to compute the test statistic (4.123), which will have an asymptotic $\chi^2_{(g_1)}$ distribution under H_0.

There are many different versions of this test, depending upon the estimates for the disturbance variance in (4.112). No matter which estimates are plugged in, however, the test statistic will always have its prescribed asymptotic null distribution, provided only that the variance estimates are consistent under H_0.

The point is that by an appropriate choice of the variance estimate, one can obtain a test statistic with known finite sample distribution. This has been investigated in detail by *Wu* (1973).

Wu proposed four different versions of the basic test statistic, two of which have a known finite sample distribution. The one that has gained wide acceptance in empirical work uses

$$\sigma^{*2} = u^{*'}u^{*} / (T-K_1-2g_1) \tag{4.134}$$

for the test statistic (4.123), where u^{*} is the OLS residual vector from the augmented regression

$$y_1 = Y_2\beta + Z_1\gamma + (Y_2 - \hat{Y}_2)\delta + u . \tag{4.135}$$

In the present notation, the test statistic can thus be written as

$$m = T(\beta^{*} - \hat{\beta})'(\sigma^{*2}\Sigma)^{-1}(\beta^{*} - \hat{\beta}) , \tag{4.136}$$

where $\Sigma = T(\tilde{\Sigma} - \hat{\Sigma})$. Since σ^{*2} is consistent for σ^2, the test statistic (4.136) has an asymptotic $\chi^2_{(g_1)}$ distribution under H_0.

In addition, Wu shows that $T_2 = m/g_1$ has an exact F distribution with g_1 and $T-K_1-2g_1$ degrees of freedom, and can thus be used for an exact rather than an asymptotic test. In subsequent Monte Carlo experiments (Wu, 1974), he provided some evidence that this version of the test is also more powerful.

Our statistic $T_2 = m/g_1$ might look different from the T_2 in Wu (1973, p.736), but is in fact algebraically identical. This follows from Σ/T(our notation)$= (Y_2'A_2Y_2)^{-1}-(Y_2'A_1Y_2)^{-1}$ (Wu's notation), and the fact that Q_2 from Wu (1973, p.136) is the same as $u^{*'}u^{*}$ from (4.134) (*Nakamura* and *Nakamura*, 1981, p.1587).

The same modification also yields an exact version of the previous test for errors in the variables. The basic requirement in both cases is that the disturbances are normal and that all elements of X are independent of all elements of u (under H_0). However, the asymptotic versions might still be valid even when these conditions are violated. Lagged dependent variables for instance will automatically induce stochastic dependence between future values of right-hand-side variables and contemporaneous disturbances. Often (but not always, as we will see below) this will not affect the asymptotic distribution of the test statistic.

Similarly, if the disturbances are not normally distributed, OLS will no longer be asymptotically efficient, and the reasoning that has led to the basic Hausman lemma (4.102) breaks down. However, OLS remains asymptotically efficient among linear estimators, so (4.102) still applies when the comparison estimator is also linear.

An important extension of Wu's work concerns the situation where only a subset of the variables in Y_2 are tested for exogeneity. The null hypothesis here is not that the full set of stochastic regressors is uncorrelated with the disturbances, but rather that this holds for a subset of the variables in Y_2.

To make this more precise, write (4.133) in more detail as

$$y_1 = Y_2^{(1)} \beta^{(1)} + Y_2^{(2)} \beta^{(2)} + Z_1 \gamma + u_1 , \qquad (4.137)$$

where the variables in $Y_2^{(1)}$ are correlated with u_1 even under H_0, and where only the nature of the variables in $Y_2^{(2)}$ is in doubt.

The problem here is that the null hypothesis is no longer equivalent to the standard linear regression model. Consequently, OLS will no longer provide the consistent and asymptotically efficient estimates that we need for the Hausman tests. *Spencer* and *Berk* (1981) however show that a straightforward Hausman test can still be obtained by comparing two versions of a 2SLS estimate for $\beta^{(2)}$, one where $Y_2^{(2)}$ is included and one where it is excluded from the Z matrix (In fact, they initially based the test on a comparison of the estimates of the full set of regression parameters, without realizing that $V(\hat{q})$ will then be singular. This is exactly the pitfall that we have warned against above).

A Hausman test with trending data

Following established practice, we have discussed the Hausman test under the assumption (2.3), i.e.

$$\lim_{T \to \infty} (1/T) X'X = Q \quad \text{(nonsingular)} .$$

In addition we have required similar behaviour for the instruments (see 4.110). Although these assumptions are standard in large sample asymptotics, it is worth noting that many empirically important situations such as trending data are thereby excluded. If for instance $x_t = t$, we have

$$X'X = \Sigma_{t=1}^{T} x_t^2 = T(T+1)(2T+1)/6 = O(T^3) , \qquad (4.138)$$

so obviously (2.3) can no longer hold.

Krämer (1984a, 1984b, 1985a, 1985d) has shown that many well known large sample results do no longer hold when regressors behave as in (4.138). In particular, OLS can be consistent, and have the same limiting distribution as competing estimators, even in the presence of simultaneous equations or errors in variables.

This raises the following questions concerning the large sample asymptotics of the Hausman test: (i) will the test statistic under H_0 still have an asymptotic χ^2 distribution when there is a trend in the data , and (ii) can we hope that the Hausman test will remain consistent in the face of identical limiting distributions, under the alternative, of $\tilde{\beta}$ and $\hat{\beta}$?

The remainder of this subsection adresses these issues for a simple Hausman test for errors in variables, with the reassuring result that both questions can be answered in the affirmative. Although this is no general proof that trending data leave Hausman tests intact, it shows at least for a special case that trend need not by necessity do any harm either.

To avoid unrelated complications that arise from the possible singularity of $V(\hat{q})$, we specialize the general errors-in-variables model (4.106) to the case of a single independent variable $\tilde{x}_t$, i.e.

$$y_t = \beta \tilde{x}_t + \epsilon_t \quad (t=1,...,T) , \tag{4.139}$$

where ϵ_t is IID$(0,\sigma_\epsilon^2)$ under H_0. Under the alternative, $\tilde{x}_t$ is measured with error, i.e. the measured variable is $x_t = \tilde{x}_t + w_t$. The w_t's are also IID$(0,\sigma_w^2)$, with $E(\epsilon_t w_t) = 0$.

A convenient IV estimate for β is based on lagged values of x_t, that is,

$$\tilde{\beta} = \Sigma_{t=1}^{T} x_{t-1} y_t / \Sigma_{t=1}^{T} x_{t-1} x_t . \tag{4.140}$$

Given the equivalents of (2.3) and (4.110), that is,

$$0 < q = \lim_{T\to\infty} \frac{1}{T} \Sigma_{t=1}^{T} \tilde{x}_t^2 < \infty , \tag{4.141}$$

$$0 < c = \operatorname{plim}_{T\to\infty} \frac{1}{T} \Sigma_{t=1}^{T} x_t x_{t-1} < \infty , \tag{4.142}$$

and

$$|c| < q , \tag{4.143}$$

both $\tilde{\beta}$ and the OLS estimate $\hat{\beta}$ are consistent and asymptotically normal under H_0, with

$$V(\hat{\beta}) = \sigma_\epsilon^2 / q , \text{ and } V(\tilde{\beta}) = \sigma_\epsilon^2 q/c^2 .$$

Therefore, by the fundamental Hausman lemma (4.102) ,

$$V(\hat{q}) = V(\tilde{\beta}) - V(\hat{\beta}) = \sigma_\epsilon^2 \left(\frac{q}{c^2} - \frac{1}{q}\right) > 0 , \tag{4.144}$$

and the test statistic (4.101) has an asymptotic $\chi^2_{(1)}$ distribution.

Under the alternative, $\operatorname{plim} \tilde{\beta} = \beta$, whereas

$$\operatorname{plim}_{T\to\infty} \hat{\beta} = \beta - \beta \frac{\sigma_w^2}{q+\sigma_w^2} \neq \beta \tag{4.145}$$

in view of the general formula (4.108).

Clearly this implies $\operatorname{plim} \hat{q} \neq 0$ and $m \xrightarrow{P} \infty$ (provided only that $1/\hat{V}(\hat{q})$ remains stochastically bounded also under the alternative, which is not a very strong requirement), so the Hausman test is consistent.

Now consider the case where there is a trend in x_t. For simplicity, let $\bar{x}_t = t$. Then (2.3) can no longer hold in view of (4.183). By theorem 2 in *Krämer* (1985d), both $\sqrt{T^3}(\tilde{\beta}-\beta)$ and $\sqrt{T^3}(\hat{\beta}-\beta)$ now have a limiting $N(0,3\sigma_\epsilon^2)$ distribution, whether there is measurement error in $\bar{x}_t$ or not.

It is easily seen that this does not affect the asymptotic $\chi^2_{(1)}$ distribution of the test statistic m from (4.101), provided $V(\hat{q})$ is estimated as usual by

$$\hat{V}(\hat{q}) = T \; \hat{\sigma}^2 \left\{ \Sigma \bar{x}_{t-1}^2 \; / \; (\Sigma \bar{x}_t \bar{x}_{t-1})^2 - 1/\Sigma \bar{x}_t^2 \right\}, \tag{4.146}$$

where $\hat{\sigma}^2$ is an arbitrary consistent estimate (under H_0) for σ_ϵ^2. This is trivially so when the ϵ_t's are normal and can otherwise be inferred from theorem 1 in *Krämer* (1985d).

More generally, write $\hat{q}$ as

$$\hat{q} = A \, y$$

with some $K \times T$ matrix A. By theorem 1 in Krämer (1985d), a sufficient condition for $\hat{q}$ to be asymptotically normal is that the maximum diagonal element of $A(A'A)A'$ tends to zero in probability as $T \to \infty$. This will always hold under standard assumptions, but also for many types of trended data. Exceptions are exponential types of trend, where one can show by simple counterexamples (take for instance $\bar{x}_t = 2^t$) that $\hat{q}$ will no longer be asymptotically normal, after proper normalization, unless the disturbances ϵ_t themselves are normal. In such cases, there is no hope to establish an asymptotic χ^2 distribution for the test statistic m.

After several lines of straightforward but tedious algebra, (4.146) can in the present simple case be written as

$$\hat{V}(\hat{q}) = \hat{\sigma}^2 \, 3T^3 \, (T+1)(T-1) \, / \, \Sigma t^2 \, (\Sigma t(t-1))^2, \tag{4.147}$$

which is $O_p(T^{-4})$ and thus of smaller order in probability than $\hat{V}(\tilde{\beta})$ and $\hat{V}(\hat{\beta})$, which are both $O_p(T^{-3})$. Contrary to trendless data, therefore, $V(\hat{q}) \neq V(\tilde{\beta}) - V(\hat{\beta})$ if identical normalizing factors are used.

Now consider the power of the test. If $\sigma_w^2 > 0$, it is easily checked that

$$\hat{q} = -\Sigma_{t=1}^T \epsilon_t \, w_t \, / \, \Sigma_{t=1}^T x_t^2 + O_p(T^{-2}), \tag{4.148}$$

where $T^2 \Sigma \epsilon_t w_t / \Sigma x_t^2 \xrightarrow{P} -3\beta\sigma_w^2 > 0$. Since $\hat{V}(\hat{q})$ is still as in (4.146), plus terms which are $O_p(T^{-4})$, we have $m \xrightarrow{P} \infty$ and a consistent test. This is so despite the asymptotic equivalence of the two estimators, and follows from the rapid convergence to zero of $\hat{V}(\hat{q})$, which outpaces the squared difference in the estimates.

Hausman tests versus classical tests

Although Hausman tests are derived from a novel idea, they look rather familiar if viewed from a different angle. To see this, reconsider the RESET test from section 4.b, where we had augmented our basic regression model (2.2) by some $T \times p$ matrix Z to obtain

$$y = X\beta + Z\gamma + u \,, \tag{4.149}$$

where $\gamma = 0$ under H_0 and $\gamma \neq 0$ under the alternative.

We will see in chapter 5 below that most alternatives to the standard regression model (2.2) can be put like this, where (4.149) obeys the classical assumptions either exactly or approximately (in a sense to be specified below). In particular, the alternatives for many Hausman tests can be written as in (4.149).

Classical procedures like RESET (which is simply the familiar F-test for omitted variables, as we have noted above) then determine whether or not the OLS estimate $\hat{\gamma}$ for γ differs significantly from zero. Hausman tests, on the other hand, compare two different estimates $\tilde{\beta}$ and $\hat{\beta}$ for β, one that is obtained by applying OLS to (4.149) ($\tilde{\beta}$) and one that is obtained after omitting Z ($\hat{\beta}$). The restricted OLS estimator $\hat{\beta}$ will be consistent and efficient under H_0, but will be inconsistent under H_1. The comparison estimator $\tilde{\beta}$ will be consistent both under H_0 and under the alternative.

As we have discussed at length above, the Hausman test rejects H_0 whenever $\hat{q} = \tilde{\beta} - \hat{\beta}$ is too large. In the present context, the covariance matrix of $\hat{q}$ can be written as

$$V(\hat{q}) = \sigma^2 \left[(X'Q_zX)^{-1} - (X'X)^{-1} \right] \,, \tag{4.150}$$

where $Q_z = I - P_z = I - Z(Z'Z)^{-1}Z'$, so one version of the test statistic is

$$m = (T-K-p)\hat{q}' \left[(X'Q_zX)^{-1} - (X'X)^{-1} \right]^{+} \hat{q} / \tilde{u}'\tilde{u} \,, \tag{4.151}$$

where $\tilde{u} = y - X\tilde{\beta} - Z\tilde{\gamma}$ is the residual vector from applying OLS to (4.149), and $[\ ]^{+}$ denotes the Moore-Penrose inverse.

By contrast, the F-test of H_0: $\gamma = 0$ rejects for large values of

$$F = \tilde{\gamma}'\tilde{V}(\tilde{\gamma})^{-1}\tilde{\gamma} / p \tag{4.152}$$

where

$$\tilde{\gamma} = (Z'Q_xZ)^{-1} Z'Q_xy \tag{4.153}$$

is the OLS estimate for γ in (4.149),

$$\tilde{V}(\tilde{\gamma}) = \frac{\tilde{u}'\tilde{u}}{T-K-P} (Z'Q_xZ)^{-1} \,, \tag{4.154}$$

and $Q_x = I - P_x = I - X(X'X)^{-1}X'$.

The obvious question is: how do the m and F tests compare?

We examine this problem under the assumption that $X'Z$ has full rank, i.e that its ranks equals min(K,p). This is not a strong requirement, since Z is chosen in many applications to resemble X, so the relationship between these matrices should be as far away as possible from orthogonality. Also, remember that the disturbances u_t are always taken to be normal, unless explicitly stated otherwise. Finally, we take X and Z to be nonstochastic.

The latter assumptions concern only the stochastic properties of the tests. They are irrelevant for the arithmetic relationships between the respective test statistics, which hold irrespective of the type of the regressors in the model.

Given this set-up, we first consider the case when $K \geq p$.

The main result is that when m is computed as in (4.151) and $X'Z$ has rank p, *the Hausman and F-tests are identical*. By this we mean that the algebraic identity

$$F = m/p \qquad (4.155)$$

holds. When the disturbances are normal and when the regressors are nonstochastic, the asymptotic Hausman test can thus be made exact by noting that m/p has under H_0 an F distribution with p and T-K-p degrees of freedom. Its rejection region, if derived from this result, is identical to the rejection region for the classical F-test.

For proof of (4.155), rewrite the F-statistic (4.152), using (4.154), as

$$F = \frac{(T-K-p)}{p} \, \tilde{\gamma}'(Z'Q_xZ)\tilde{\gamma}/\tilde{u}'\tilde{u} , \qquad (4.156)$$

and note the following useful relationship between $\hat{q}$ and $\tilde{\gamma}$:

$$\hat{q} = \tilde{\beta} - \hat{\beta} = -(X'X)^{-1} X'Z\tilde{\gamma} . \qquad (4.157)$$

This follows for instance from

$$\begin{bmatrix} X'X & X'Z \\ Z'X & Z'Z \end{bmatrix} \begin{bmatrix} \tilde{\beta} \\ \tilde{\gamma} \end{bmatrix} = \begin{bmatrix} X'y \\ Z'y \end{bmatrix}, \qquad (4.158)$$

which in turn implies

$$(X'X)\tilde{\beta} + (X'Z)\tilde{\gamma} = X'y \qquad (4.159)$$

and

$$\begin{aligned} \tilde{\beta} &= -(X'X)^{-1} X'Z\tilde{\gamma} + (X'X)^{-1} X'y \\ &= -(X'X)^{-1} X'Z\tilde{\gamma} + \hat{\beta} . \end{aligned} \qquad (4.160)$$

The identity (4.155) now follows from the following identity, which can be derived from *Rao* and *Mitra* (1971, pp.70-71) (see also *Hausman* and *Taylor*, 1980, lemma 2.4):

$$(X'Q_zX)^{-1} = (X'X)^{-1} + (X'X)^{-1} X'Z(Z'Q_xZ)(Z'X) (X'X)^{-1} . \qquad (4.161)$$

Together, the relationships (4.157) and (4.161) imply that the test statistic (4.151) for the Hausman test can be written as

$$m = (T-K-p) \, \tilde{\gamma}' \, \{Z'X(X'X)^{-1}[(X'X)^{-1}X'Z(Z'Q_xZ)^{-1}Z'X(X'X)^{-1}]^+ (X'X)^{-1}X'Z\} \tilde{\gamma}/\tilde{u}'\tilde{u}, \qquad (4.162)$$

and (4.155) follows from the immediate fact that the matrix in braces equals $Z'Q_xZ$.

The result (4.155) shows that it does not matter, given the model can be cast under the alternative in the form (4.149), and given that Z does not have more columns than X, whether we apply a Hausman test or stay with the more conventional F-test. When we use version (4.151) of the Hausman test, the respective rejection regions will be identical.

Consider next the case $K < p$. Here the Hausman and F-tests will in general differ.

The F-statistic (4.156) continues to be distributed as $F_{p,T-K-p}$ under H_0, whereas the null distribution of m/p from (4.151) is now $F_{K,T-K-p}$. Both facts are straightforwardly verified from the algebraic expressions for the test statistics. Consequently the rejection regions for the tests will no longer coincide, and the question arises which of the tests has higher power.

Under the alternative, F and m/p have a non-central F-distribution. The non-centrality parameter, which applies only to the numerator, depends on γ, X and Z, and is derived in *Hausman* and *Taylor* (1980). Obviously it is the same for both tests when $K \geq p$. When $K < p$, the non-centrality parameter of the m-test is less than or equal to that of the F-test (Hausman and Taylor, 1980, proposition 2.7). Since the rejection probability for any given size is an increasing function of the non-centrality parameter, this implies, ceteris paribus, that the F-test has higher power. On the other hand, m has fewer degrees of freedom than F, which counteracts this effect, since the power of the F-test decreases with the degrees of freedom of the numerator for fixed non-centrality parameter. No test is therefore uniformly superior to the other over the whole parameter space.

There are some alternatives against which the Hausman m-test performs particularly poorly. It is easy to see why. When $K < p$, there are non-zero values for γ that still imply $E(\hat{q}) = 0$. From (4.157), we have

$$E(\hat{q}) = -(X'X)^{-1} X'Z\gamma . \qquad (4.163)$$

When $K < p$, the null space of $X'Z$ is non-empty. Thus, when $\gamma \neq 0$ lies in that space, $E(\hat{q}) = 0$ also under H_1. Consequently, the distribution of the test statistic is the same as under H_0 and the power equals the size of the test.

This leads to the following important point. Whenever $E(\hat{q}) = 0$, we have $E(\hat{\beta}) = \beta$. In particular, the restricted OLS estimate $\hat{\beta}$ can be unbiased, and under the present assumptions also consistent, even under the alternative $\gamma \neq 0$. In view of (4.163), the only requirement is

$$(X'X)^{-1} X'Z\gamma = 0 . \qquad (4.164)$$

For $K \geq p$, this holds if and only if $\gamma = 0$, but for $K < p$, the condition $\gamma = 0$ is no longer necessary. The equality (4.164) can occur even when $\gamma \neq 0$. Thus, when the aim of the specification analysis is to examine whether or not OLS produces unbiased and consistent estimates for the parameters in (4.149), the appropriate null hypothesis is (4.164) rather than H_0: $\gamma = 0$. The former null hypothesis is always implicitly used by the Hausman test, which thus examines the effect of imposing the restrictions themselves in the light of the data.

Of course, (4.164) can also be subjected to a conventional F-test. Calling the resulting statistic F^K, one can show along the same lines as above that then again $m/p = F^K$, i.e. the Hausman test and the conventional F-test of the modified null hypothesis (4.164) produce identical results.

The differencing test

A straightforward application of Hausman's basic idea leads to the differencing test, which compares the OLS estimates to the First Difference (FD) estimates of β. This procedure has long been informally used in empirical work, where very often the data are used both in levels and in differenced form. Whenever the resulting coefficient estimates deviate too much, there is certainly reason to believe that something is wrong with the model - a very intuitive idea that has recently been formalized by *Plosser, Schwert and White* (1982). Related contributions are *Davidson* and *MacKinnon* (1984), *Breusch* and *Pagan* (1984), *Godfrey* (1984), and *Davidson, Godfrey* and *MacKinnon* (1986).

Consider our standard regression model $y = X\beta + u$ and also the differenced regression

$$\dot{y} = \dot{X}\beta + \dot{u},$$ (4.165)

where $\dot{y} = Dy$, $\dot{X} = DX$, $\dot{u} = Du$ and where

$$D = \begin{bmatrix} 1 & -1 & & 0 \\ & 1 & -1 & \\ 0 & & 1 & -1 \end{bmatrix}$$ (4.166)

is the familiar $(T-1) \times T$ differencing matrix.

Whenever there is a constant in the initial regression, the first column of X becomes zero and is dropped. Therefore we assume for simplicity that there is no constant in the levels regression, and that X has full column rank (a constant merely complicates the notation and can therefore without loss of generality be assumed away).

The FD estimate for β is then

$$\tilde{\beta} = (\dot{X}'\dot{X})^{-1}\dot{X}'\dot{y},$$ (4.167)

with covariance matrix

$$cov(\tilde{\beta}) = \sigma^2(\dot{X}'\dot{X})^{-1}\dot{X}'DD'\dot{X}(\dot{X}'\dot{X})^{-1},$$ (4.168)

where

$$DD' = \begin{bmatrix} 2 & -1 & & 0 \\ -1 & 2 & & -1 \\ 0 & & -1 & 2 \end{bmatrix}$$ (4.169)

is the covariance matrix of the disturbance vector u from the differenced regression (4.165).

The OLS estimate $\hat{\beta}$ is uncorrelated with $\tilde{\beta} - \hat{\beta}$, since both OLS and FD are linear, and OLS is BLUE. Thus, the difference $\hat{q} = \tilde{\beta} - \hat{\beta}$ has covariance matrix

$$cov(\hat{q}) = \sigma^2 [cov(\tilde{\beta}) - cov(\hat{\beta})].$$ (4.170)

If in addition

$$\frac{1}{T}\dot{X}'\dot{X} \to \dot{Q} \quad \text{(nonsingular)},$$ (4.171)

$\sqrt{T}\hat{q}$ has a limiting normal distribution with covariance matrix

$$V(\hat{q}) = V(\tilde{\beta}) - V(\hat{\beta}) ,$$

where $V(\tilde{\beta})$ and $V(\hat{\beta})$ are the covariance matrices of the limiting distributions of $\sqrt{T}(\tilde{\beta}\text{-}\beta)$ and $\sqrt{T}(\hat{\beta}\text{-}\beta)$, respectively. A consistent estimate for $V(\hat{q})$ is

$$\hat{V}(\hat{q}) = \hat{\sigma}^2[(\dot{X}'\dot{X}/T)^{-1} (\dot{X}'DD'\dot{X}/T) (\dot{X}'\dot{X}/T)^{-1}- (X'X/T)^{-1}] \tag{4.172}$$

where $\hat{\sigma}^2$ is any consistent estimate for σ^2.

The test statistic for the differencing test is

$$\Delta = T\hat{q}' \hat{V}(\hat{q})^{-1}\hat{q} , \tag{4.173}$$

where we implicitly assume $\hat{V}(\hat{q})$ nonsingular. The null hypothesis that the initial levels model is correctly specified is rejected whenever Δ is too large. When in addition $V(\hat{q})$ is nonsingular, Δ has an asymptotic $\chi^2_{(K)}$ distribution under H_0.

This leads to the first problem with the differencing test. Often $\text{cov}(\hat{q})$ and $V(\hat{q})$ are singular. This occurs for instance in polynomial regression or in distributed lag models, as can be shown by simple examples.

The obvious solution is a generalized inverse $\hat{V}(\hat{q})^-$ in (4.173). Similar to the general Hausman test, this is guaranteed to work only if $\hat{V}(\hat{q})$ has the same rank as $V(\hat{q})$ with a probability that tends to one. Then Δ has an asymptotic χ^2 distribution with K' rather than K degrees of freedom, where $K' = \text{rank } V(\hat{q})$.

Similar again to the Hausman test, there are systematic forces at work which prevent any nonsingularity of $V(\hat{q})$ from being detected. In practice the matrix $\dot{X}' DD'\dot{X}$ from the expressions (4.169) and (4.172) is often replaced by the simpler expression $\ddot{X}'\ddot{X}$, where $\ddot{X}$ is the X-matrix differenced twice, i.e. $\ddot{x}_t = \dot{x}_t - \dot{x}_{t-1}$ $(t = 3,...,T)$. Since

$$\frac{1}{T}[\dot{X}'DD'\dot{X} - \ddot{X}'\ddot{X}] = \frac{1}{T}[\dot{x}_T\dot{x}_T' + \dot{x}_2\dot{x}_2'] \rightarrow 0 , \tag{4.174}$$

the consistency of the estimate for $V(\hat{q})$ is not endangered by this simplification. If we denote the modified estimate by $\tilde{V}(\hat{q})$, (4.174) implies

$$\tilde{V}(\hat{q}) - \hat{V}(\hat{q}) \xrightarrow{p} 0 ,$$

so $\tilde{V}(\hat{q}) \xrightarrow{p} V(\hat{q})$ whenever $\hat{V}(\hat{q}) \xrightarrow{p} V(\hat{q})$. However, there is no guarantee that $\tilde{V}(\hat{q})^-$ has the same probability limit as $\hat{V}(\hat{q})^-$. This will happen only when $\tilde{V}(\hat{q})$ and $\hat{V}(\hat{q})$ have the same rank, which need not to be the case. Also the limiting distribution of the test statistic might now be more seriously affected than in the case of the Hausman test for errors in variables. We do not investigate this matter any further, since we will see below that there are other ways to compute the test statistic (4.172) which circumvent these problems.

Another drawback of the differencing test is that it does not generalize easily to dynamic models. The disturbances in the differenced regression have covariance matrix (4.168) and are thus autocorrelated (they follow an MA(1) process by construction). Thus lagged dependent variables are correlated with the contemporaneous disturbances and OLS will be inconsistent *also under* H_0. There is then no hope to establish the asymptotic null distribution of the test statistic (4.172).

The differencing test is consistent whenever

$$\text{plim}_{T-\infty} \tilde{\beta} \neq \text{plim}_{T-\infty} \hat{\beta} \qquad (4.175)$$

under the alternative, since the test statistic Δ will then tend to infinity in probability. Unlike many other Hausman tests, the estimate $\tilde{\beta}$ is here not chosen with any particular alternative in mind - we have a pure misspecification test or significance test. This label however does not mean much, since Hausman tests in general, whether the comparison estimator is chosen with a view towards a particular alternative or not, have power whenever (4.175) holds. The requirement imposed in Hausman (1978) that $\hat{\beta}$ be consistent under the alternative, which in conjunction with the inconsistency of OLS implies (4.175), is not needed for this test to work.

First differences as an IV estimator

The differencing test, as a special Hausman test, uses for comparison the OLS coefficient estimate from the transformed data

$$\dot{y}_t = y_t - y_{t-1}, \dot{x}_t = x_t - x_{t-1} \quad (t=2,...,T).$$

A natural extension are "pseudo-differences"

$$\dot{y}_t = y_t - \varrho y_{t-1}, \dot{x}_t = x_t - \varrho x_{t-1} \quad (t=2,...,T) \qquad (4.176)$$

which are often used to eliminate first order serial correlation in the disturbances. OLS estimation subsequent to the transformation (4.176) will likewise produce consistent etimates when the model is correct and will lead to a different probability limit than straightforward OLS otherwise. The question then is: Is there any claim that $\varrho = 1$ will produce the best test, in any sense, and if not, which value for ϱ is optimal?

The answer is that the choice of ϱ does not matter much. Given an appropriate estimate for the covariance matrix of the difference of the estimates, the test statistics will be almost identical.

This important result was independently discovered by several authors, among them *Breusch* and *Pagan* (1984) and *Davidson, Godfrey* and *MacKinnon* (1986). The proof is in two steps. First we show that the First Differences estimator β is in fact identical to a certain Instrumental Variables (IV) estimator, given an appropriate choice of the instruments.

Any IV estimator of the model $y = X\beta + u$ can be written as

$$\tilde{\beta} = (H'X)^{-1} H'y, \qquad (4.177)$$

where the columns of the $T \times K$ matrix H are the instruments. Depending upon H, there are many possible IV estimators. We have met various members of this species before. The point here is that given a proper choice of H, the expressions (4.177) and (4.167) are identical.

To see this, define the t'th row of H as follows $(t = 2,...,T-1)$:

$$h_t = 2x_t - x_{t+1} - x_{t-1} . \tag{4.178}$$

The remaining rows are

$$h_1 = x_1 - x_2 , \quad h_T = x_T - x_{T-1} . \tag{4.179}$$

From (4.178), it is clear that H is essentially equal to $\ddot{X}_{+1}$, where "$+1$" means that the t'th row of H equals the $t+1$'st row of $\ddot{X}$. The equality is exact if

$$x_{T+1} = x_T \text{ and } x_0 = x_1 , \tag{4.180}$$

since h_1 and h_T are then also determined from (4.178).

The identity $\tilde{\beta}(\text{FD}) = \tilde{\beta}(\text{IV})$ now follows from

$$\dot{X}'\dot{X} = H'X \tag{4.181}$$

and

$$\dot{X}'\dot{y} = H'y . \tag{4.182}$$

For proof of (4.181), let R denote any column of X, and let S denote the same or another column of X. Then

$$
\begin{aligned}
\dot{R}'\dot{S} &= \Sigma_{t=2}^{T} (r_t - r_{t-1})(s_t - s_{t-1}) \\
&= \Sigma_{t=2}^{T} r_t s_t + \Sigma_{t=2}^{T} r_{t-1}s_{t-1} - \Sigma_{t=2}^{T} r_t s_{t-1} - \Sigma_{t=2}^{T} r_{t-1}s_t \\
&= 2 \Sigma_{t=2}^{T-1} r_t s_t - \Sigma_{t=2}^{T} r_t s_{t-1} - \Sigma_{t=2}^{T} r_{t-1}s_t + r_1 s_1 + r_T s_T.
\end{aligned}
\tag{4.183}
$$

The corresponding entry in the matrix H'X is similarly computed as

$$
\begin{aligned}
&\Sigma_{t=2}^{T-1} (2r_t - r_{t+1} - r_{t-1})s_t + (r_1 - r_2)s_1 + (r_T - r_{T-1})s_T \\
&= 2 \Sigma_{t=2}^{T-1} r_t s_t - \Sigma_{t=2}^{T-1} r_{t+1}s_t - \Sigma_{t=2}^{T-1} r_{t-1}s_t + r_1 s_1 - r_2 s_1 + r_T s_T - r_{T-1}s_T ,
\end{aligned}
\tag{4.184}
$$

which is the same as (4.183). This shows that $\dot{X}'\dot{X} = H'X$, when H is defined as in (4.178) and (4.179). The relationship (4.182) is established in the same way. Together, this shows that FD is indeed identical to a certain version of the IV method.

The same is along identical lines also shown for estimates based on the pseudo-differences (4.176). Here, H is defined by

$$h_t = (1 + \varrho^2) x_t - \varrho x_{t+1} - \varrho x_{t-1} , \quad (t = 2,...,T-1) \tag{4.185}$$

$$h_1 = \varrho^2 x_1 - \varrho x_2 , \tag{4.186}$$

and

$$h_T = x_T - \varrho x_{T-1} . \tag{4.187}$$

Obviously, this amounts to (4.178) and (4.179) in the FD case where $\varrho = 1$. (4.185) now applies to all rows of H if

$$x_{T+1} = \varrho x_T \text{ and } x_0 = \frac{1}{\varrho} x_1 . \tag{4.188}$$

Again, this leads to the analoguous expression (4.180) when $\varrho = 1$.

The near equality of Hausman tests based on the differences of the initial data is now shown in a second step by establishing that any Hausman test that compares OLS to an IV estimate can be viewed as an F-test for H_0: $\gamma = 0$ in the equation $y = X\beta + Z\gamma + u$. The Z-matrix there of course varies depending upon which IV estimate is used, but it is almost identical for differenced and pseudo-differenced data. We will return to this point in more detail in chapter 5 below.

The information matrix test

So far, Hausman-type specification tests were based on different estimates of the regression coefficients. However, the same idea can also be applied to parameters other than those of primary interest. One such set of secondary parameters is given by the entries of the *information matrix*.

The information matrix $I(\theta)$ was introduced in (2.22) above for an arbitrary statistical model. Usually, it depends on the unknown parameters in θ and is thus itself unknown. By the fundamental equality (2.42), there are at least two different ways of estimating $I(\theta)$ in a given application. One is to evaluate the expectation matrix of second derivatives of the log likelihood at the ML estimate $\hat\theta$, and the other is to sum up the outer products of the score vectors as in (2.45). These estimates should be close to each other when the model is correct and can thus be used as the basis of another Hausman-type specification test. *White* (1982) has formalized this idea for the general case. Our exposition follows *Hall* (1983), where White's general procedure is specialized to the standard linear regression model.

In the linear regression model $y = X\beta + u$, let $\theta = [\beta', \sigma^2]'$ denote the vector of unknown parameters. The log likelihood of the t'th observation is then

$$L(\theta, y_t) = -\frac{1}{2} \log 2\pi - \frac{1}{2} \log \sigma^2 - \frac{1}{2\sigma^2} (y_t - x_t'\beta)^2 . \tag{4.189}$$

The first order partial derivatives with respect to β and σ^2 are (compare 2.28)

$$\frac{\partial L}{\partial \beta} = \frac{1}{\sigma^2} (y_t - x_t'\beta) x_t \tag{4.190}$$

and

$$\frac{\partial L}{\partial \sigma^2} = -\frac{1}{2\sigma^2} + \frac{1}{2\sigma^4} (y_t - x_t'\beta)^2 . \tag{4.191}$$

The second order partial derivatives are

$$\frac{\partial^2 L(y_t, \theta)}{\partial \beta \partial \beta'} = -\frac{1}{\sigma^2} x_t x_t' , \tag{4.192}$$

$$\frac{\partial^2 L(y_t, \theta)}{\partial \beta \partial \sigma^2} = -\frac{1}{\sigma^4} u_t x_t , \tag{4.193}$$

and

$$\frac{\partial^2 L(y_t,\theta)}{(\partial\sigma^2)^2} = \frac{1}{2\sigma^4} - \frac{u_t^2}{\sigma^6}. \tag{4.194}$$

The information matrix of the full sample is therefore

$$I(\theta) = E A(\theta), \tag{4.195}$$

where

$$A(\theta) = -\frac{\partial^2 L(y,\theta)}{\partial\theta\partial\theta'}$$

$$= -\Sigma_{t=1}^{T} \frac{\partial^2 L(y_t,\theta)}{\partial\theta\partial\theta'}$$

$$= \Sigma_{t=1}^{T} \begin{bmatrix} \frac{1}{\sigma^2} x_t x_t' & \frac{1}{\sigma^4} u_t x_t \\ \frac{1}{\sigma^4} u_t x_t' & -\frac{1}{2\sigma^4} + \frac{u_t^2}{\sigma^6} \end{bmatrix} \tag{4.196}$$

which implies

$$I(\theta) = \begin{bmatrix} \frac{1}{\sigma^2} X'X & 0 \\ 0 & \frac{T}{2\sigma^4} \end{bmatrix}. \tag{4.197}$$

Given assumption (2.3) (i.e. $X'X/T \to Q$ for some nonsingular and finite matrix Q), $I(\theta)$ converges to some finite nonsingular matrix $\bar{I}$, after division by the sample size T:

$$\bar{I} = \lim_{T\to\infty} T^{-1} I(\theta) = \begin{bmatrix} \frac{1}{\sigma^2} Q & 0 \\ 0 & \frac{1}{2\sigma^4} \end{bmatrix}. \tag{4.198}$$

By the basic relationship (2.24), $I(\theta)$ can also be expressed as

$$I(\theta) = E B(\theta), \tag{4.199}$$

where

$$B(\theta) = \Sigma_{t=1}^{T} \left(\frac{\partial L(y_t,\theta)}{\partial\theta}\right) \left(\frac{\partial L(y_t,\theta)}{\partial\theta}\right)'$$

$$= \Sigma_{t=1}^{T} \begin{bmatrix} \frac{u_t^2}{\sigma^4} x_t x_t' & -\frac{u_t x_t}{2\sigma^4} + \frac{u_t^3 x_t}{2\sigma^6} \\ -\frac{u_t x_t'}{2\sigma^4} + \frac{u_t^3 x_t'}{2\sigma^6} & \frac{1}{4\sigma^4} - \frac{u_t^2}{2\sigma^6} + \frac{u_t^4}{4\sigma^8} \end{bmatrix} \tag{4.200}$$

Both (4.195) and (4.199) can be used to estimate the unknown matrix $\bar{I}$. This is done by replacing β, σ^2 and u in (4.196) and (4.200) with their respective maximum likelihood estimates $\hat{\beta} = (X'X)^{-1}X'y$, $\hat{\sigma}^2 = \hat{u}'\hat{u}/T$ and $\hat{u} = y - X\hat{\beta}$. If we denote the resulting matrices by $\hat{A}(\theta)$ and $\hat{B}(\theta)$, respectively, we have

$$\hat{A}(\theta) = \begin{bmatrix} \dfrac{1}{\hat{\sigma}^2} X'X & 0 \\[2ex] 0 & \dfrac{T}{2\hat{\sigma}^4} \end{bmatrix} \tag{4.201}$$

and

$$\hat{B}(\theta) = \begin{bmatrix} \dfrac{1}{\hat{\sigma}^4} \Sigma \hat{u}_t^2 x_t x_t' & \dfrac{1}{2\hat{\sigma}^6} \Sigma \hat{u}_t^3 x_t' \\[3ex] \dfrac{1}{2\hat{\sigma}^6} \Sigma \hat{u}_t^3 x_t' & -\dfrac{T}{4\hat{\sigma}^4} + \dfrac{\Sigma \hat{u}_t^4}{4\hat{\sigma}^8} \end{bmatrix} \tag{4.202}$$

If the model is correctly specified, including normality of the disturbances u_t, and given the large sample assumption (2.3), it is straightforwardly shown that

$$\text{plim}_{T\to\infty} \frac{1}{T}\hat{A}(\theta) = \text{plim}_{T\to\infty} \frac{1}{T}\hat{B}(\theta) = \tilde{I}. \tag{4.203}$$

The information matrix test therefore rejects the model whenever

$$\frac{1}{T}[\hat{B}(\theta) - \hat{A}(\theta)] \tag{4.204}$$

is too large.

The only problem is the practical implementation of this simple idea. The matrix (4.204) has $(K+1) \times (K+1)$ elements, of which only $(K+2)(K+1)/2$ are different, due to the symmetry of the matrix. We therefore confine ourselves to the $(K+2)(K+1)/2$ dimensional indicator vector $\hat{D}(\theta)$ whose elements consist of the lower triangular elements of (4.204).

$\hat{D}(\theta)$ is composed of three subvectors. The first $K(K+1)/2$ elements of $\hat{D}(\theta)$ are the subvector

$$\Delta_1 = \frac{1}{T\hat{\sigma}^4} \Sigma_{t=1}^{T} \begin{bmatrix} (\hat{u}_t^2 - \hat{\sigma}^2) x_{1t}^2 \\ (\hat{u}_t^2 - \hat{\sigma}^2) x_{2t} x_{1t} \\ \vdots \\ (\hat{u}_t^2 - \hat{\sigma}^2) x_{Kt}^2 \end{bmatrix} \tag{4.205}$$

The s'th element of Δ_1 is $\dfrac{1}{T\hat{\sigma}^4} \Sigma_{t=1}^{T} (\hat{u}_t^2 - \hat{\sigma}^2) x_{it} x_{jt}$,

where $s = (j-1)(K-j/2) + i$; $i \geq j$; $i,j = 1,...,K$. This subvector measures the discrepancy between two estimates of the covariance matrix of $\hat{\beta}$, similar to White's heteroskedasticity test from section 3.b.

The next K elements of the indicator vector $\hat{D}(\theta)$ are

$$\Delta_2 = \frac{1}{2T\hat{\sigma}^6} \Sigma_{t=1}^{T} \hat{u}_t^3 x_t'. \tag{4.206}$$

This equals the last row of $\hat{B}(\theta)$ except the last element and measures the discrepancy of the estimates of the covariance of $\hat{\beta}$ and $\hat{\sigma}^2$.

The last element of $\hat{D}(\theta)$ is the difference of the bottom right elements of $\hat{B}(\theta)/T$ and $\hat{A}(\theta)/T$, i.e.

$$\Delta_3 = -\frac{3}{4\hat{\sigma}^4} + \frac{1}{T}\Sigma_{t=1}^T \frac{\hat{u}_t^4}{4\hat{\sigma}^8}. \tag{4.207}$$

This compares the respective estimates of the variance of $\hat{\sigma}^2$.

The complete indicator vector can in this notation be expressed as

$$\hat{D}(\theta) = [\Delta_1', \Delta_2', \Delta_3]'.$$

From White (1982), $\sqrt{T}\hat{D}(\theta)$ has under H_0 a normal limiting distribution with mean zero and some covariance matrix $V(\theta)$. If $V(\theta)$ is nonsingular, the appropriate test statistic is

$$\xi = T\,\hat{D}(\theta)'\,V(\theta)^{-1}\hat{D}(\theta),$$

which has a limiting χ^2 distribution with $(K+2)(K+1)/2$ degrees of freedom under H_0.

In most applications, $V(\theta)$ is unknown and must be estimated. In the present context, a consistent estimate for $V(\theta)$ is obtained from the sample moments of $\hat{D}(\theta)$:

$$\tilde{V}(\theta) = \frac{1}{T}\Sigma_{t=1}^T \begin{bmatrix} \Delta_{1t}\Delta_{1t}' & \cdot & \cdot \\ \Delta_{2t}\Delta_{1t}' & \Delta_{2t}\Delta_{2t}' & \cdot \\ \Delta_{3t}\Delta_{1t}' & \Delta_{3t}\Delta_{2t}' & \Delta_{3t}^2 \end{bmatrix}. \tag{4.208}$$

Since in addition the off-diagonal blocks of $V(\theta)$ can be shown to be zero (see the appendix of Hall, 1983), we can drop the corresponding off-diagonal elements of $\tilde{V}(\theta)$ to obtain our final estimate

$$\hat{V}(\theta) = \begin{bmatrix} \Delta^{11} & 0 & 0 \\ 0 & \Delta^{22} & 0 \\ 0 & 0 & \Delta^{33} \end{bmatrix}, \tag{4.209}$$

where $\Delta^{ij} = \frac{1}{T}\Sigma_{t=1}^T \Delta_{it}\Delta_{jt}'$. The test statistic is thus

$$\begin{aligned} \xi &= T\,\hat{D}(\theta)'\,\hat{V}(\theta)^{-1}\hat{D}(\theta) \\ &= T\Delta_1'(\Delta^{11})^{-1}\Delta_1 + T\Delta_2'(\Delta^{22})^{-1}\Delta_2 + T\Delta_3'(\Delta^{33})^{-1}\Delta_3 \\ &= \xi_1 + \xi_2 + \xi_3, \end{aligned} \tag{4.210}$$

i.e. the sum of three asymptotically independent components. It is easily seen that ξ_1 equals a particular version of White's (1980) test for heteroskedasticity. The statistic ξ_2 can be obtained as

$$\frac{T}{6\hat{\sigma}^6}\text{ESS},$$

where ESS is the explained sum of squares from the regression of $\hat{u}_t^3$ on x_t, and $\hat{\sigma}$ is obtained from the original OLS regression of y_t on x_t. Finally, straightforward computation shows that ξ_3 equals

$$\frac{T}{24\hat{\sigma}^8}\left(\frac{1}{T}\Sigma\hat{u}_t^4 - 3\hat{\sigma}^4\right)^2,$$

which is similar to the test statistic (3.103) of the Jarque-Bera test for normality of the disturbances.

In practice, the columns of the indicator vector $\hat{D}(\theta)$ are often linear dependent. Consider for instance a model with a constant and polynomial terms, where $x_{t1} = 1$, and $x_{t3} = x_{t2}^2$. Then $x_{t1}x_{t3} = x_{t2}^2$, and the first three elements of Δ_1 are linear dependent. This also implies that $\hat{V}(\theta)$ is singular. Similar to the standard Hausman test, one could use a generalized inverse in (4.209) and reduce the degrees of freedom. An algebraically identical result is however more easily obtained by dropping any redundant terms in $\hat{D}(\theta)$ and $\hat{V}(\theta)$.

From the composition of the test statistic it is clear that the information matrix test will have power whenever the disturbances are nonnormal or heteroskedastic. As the example from section 3.b has shown, it will however also be consistent when there is misspecification among the regressors. In this sense, it can be used as a general test for misspecification.

5. UNIFYING THEMES

There are various ways to organize the many apparently unrelated procedures discussed so far. One is to group them into the Lagrange Multiplier, Wald or Likelihood Ratio classes of tests, if possible. Another illuminating similarity is that many procedures can be interpreted as either data transformation tests or data addition tests, or both.

a) Variable Transformation

A general way to generate significance tests for any model is to compare parameter estimates before and after some transformation of the observations. More precisely, choose the transformation such that, if the original model is correct, the parameters of the transformed model have a known relation to the parameters of the original model, and reject the model whenever this relationship is significantly violated by the respective estimates.

Linear transformations

In most applications, the transformation is linear, and the parameters of the original and the transformed model are identical. In addition we confine ourselves to the familiar linear regression model

$$y = X\beta + u ,$$ (5.1)

where X is $T \times K$ and where the null hypothesis maintains that assumptions A1-A4 from page 3 hold. Data transformation tests then subject the model (5.1) to a known transformation matrix F so that

$$Fy = FX\beta + Fu$$ (5.2)

or

$$y^* = X^*\beta + u^* ,$$ (5.3)

where $y^* = Fy$, $X^* = FX$, and $u^* = Fu$. The transformation matrix F is nonstochastic $n \times T$ with rank equal to n. Initially we assume in addition that $FX = X^*$ is also of rank K.

A natural idea in this framework - in fact, a straightforward application of the Hausman principle - is to compare the OLS coefficient estimates from the transformed model (5.3), i.e.

$$\beta^* = (X'F'FX)^{-1} X'F'Fy$$
$$= (X^{*'}X^*)^{-1} X^{*'}y^* \tag{5.4}$$

to the OLS estimates from the original observations, i.e. $\hat{\beta} = (X'X)^{-1}X'y$. If the original model is correct, both β^* and $\hat{\beta}$ are unbiased and consistent (given some regularity conditions about X and F). The respective covariance matrices are $cov(\hat{\beta}) = \sigma^2(X'X)^{-1}$ and

$$cov(\beta^*) = \sigma^2 (X'F'FX)^{-1} X'F'FF'F X (X'F'FX)^{-1} . \tag{5.5}$$

Since $\hat{\beta}$ is BLUE, we have

$$cov(\beta^*-\hat{\beta}) = cov(\beta^*) - cov(\hat{\beta}), \tag{5.6}$$

and the test statistic

$$m = (\beta^*-\hat{\beta})' [cov(\beta^*)-cov(\hat{\beta})]^{-1}(\beta^*-\hat{\beta}) \tag{5.7}$$

is asymptotically χ^2_K under H_0.

In practice the disturbance variance σ^2 in $cov(\beta^*)$ and $cov(\hat{\beta})$ must of course be replaced by some consistent estimate $\hat{\sigma}^2$. The test can then often be made exact by an appropriate choice of $\hat{\sigma}^2$.

Still, there are two major obstacles to this simple approach.

(i) the difference $cov(\beta^*)-cov(\hat{\beta})$ of the respective covariance matrices will often be singular, and the formula (5.7) cannot be applied as it stands. The familiar solution to this problem is a g-inverse rather than the ordinary inverse in the expression for the test statistic. However, as we have seen in chapter 4 for a special Hausman test, this requires some care, in order to guarantee the convergence of the normalized g-inverse to the g-inverse of the covariance matrix of the limiting distribution of $\sqrt{T}(\beta^*-\hat{\beta})$.

(ii) The transformed regressor matrix $X^* = FX$ may not have rank K. In such cases we simply drop some columns of X^* until the remaining regressors are linearly independent, and confine the comparison to the coefficients of the remaining regressors.

Whenever $n < T$, i.e. whenever the transformed model contains fewer observations than the original model, we may also follow an alternative strategy. Under H_0, the covariance matrix of the transformed residual vector is given by

$$cov(u^*) = \sigma^2 FF' , \tag{5.8}$$

a matrix which is known up to a constant of proportionality. It is therefore also possible to apply the generalized least squares (GLS) estimator to the transformed model (5.3) to obtain

$$\tilde{\beta} = [X^{*'}(FF')^{-1}X^*]^{-1} X^{*'}(FF')^{-1}y^* . \tag{5.9}$$

An alternative data transformation test can then be based on a comparison of $\hat{\beta}$ and $\tilde{\beta}$.

Whenever $n = T$ and F is nonsingular, we have

$$\tilde{\beta} = [X'F'(FF')^{-1}FX]^{-1}X'F'(FF')^{-1}\,Fy$$

$$= (X'X)^{-1}\,X'y$$

$$= \hat{\beta}\,,$$

i.e. OLS and GLS are identical. Since this identity also holds when F is singular, the alternative test only applies when $n < T$.

Whenever the OLS estimator for the original model can be compared to either the OLS or the GLS estimator for the transformed model, it is of course natural to ask which of the latter estimators is to be preferred, i.e. which test has higher power. Such a power comparison requires the specification of an alternative hypothesis, and will produce mixed results. Therefore no general advice can be given in this respect.

A final generalization of the data transformation approach concerns the underlying parameters. Rather than relying on the regression coefficients, one could also use the variance of the disturbances as the basis of a transformational test.

Let the t'th element of Fu be

$$u_t^* = \Sigma_{i=1}^T f_{it}\, u_i\,. \tag{5.10}$$

If the original model is correct (in particular if the u_i's are uncorrelated with common variance σ^2), the variance of u_t^* will be

$$\operatorname{var}(u_t^*) = \sigma^2 \Sigma_{i=1}^T f_{it}^2\,. \tag{5.11}$$

Since the f_{it} are known, we have the following known relationship between the variances of the transformed and the original model:

$$\frac{\Sigma\,\operatorname{var}(u_t^*)}{\Sigma\,\operatorname{var}(u_t)} = \Sigma_{i,t=1}^T f_{it}^2\,. \tag{5.12}$$

Yet another data transformation test can then be based on a check of whether or not the respective regression residuals also obey this relationship.

Examples

(i) The Plosser-Schwert-White (1982) differencing test. We have discussed this procedure at length in chapter 4 above. Here the transformation matrix F is $(T-1) \times T$ and equals the differencing matrix D from (4.166). Whenever there is a constant in the regression, the corresponding column in FX becomes zero and $FX = X^*$ has rank at most equal to $K-1$. In addition, $\operatorname{cov}(\beta^*)-\operatorname{cov}(\hat{\beta})$ can be rank deficient due to the presence of linear or polynomial trends among the regressors of the original model.

Although $m = T\text{-}1 < T$, it can be shown that OLS still equals GLS in the transformed model whenever there is a constant in the regression (see Friedman, 1977). Any comparison must therefore be based on $\hat{\beta}$ and β^*.

(ii) Farebrother's (1979) grouping test. Here F aggregates data, i.e.

$$f_{it} = \begin{cases} 1 & t\text{'th observation belongs to group } i \\ 0 & \text{otherwise} \end{cases} \qquad (5.13)$$

and m equals the number of different groups. The T original observations on any variable are thus converted to m group totals.

Farebrother (1979) has based this test on $\hat{\beta}$ and $\bar{\beta}$ (the GLS estimate from the transformed model). Recent applications include *Polinsky* (1977) or *Greenlees* and *Zieschang* (1984).

(iii) OLS compared to GLS. When the disturbance covariance matrix in the model (5.1) is given by

$$\text{cov}(u) = \sigma^2 V \text{ (nonsingular)}, \qquad (5.14)$$

it is well known that the GLS estimate $\bar{\beta}$ can be obtained by applying OLS to the transformed model

$$Ry = RX\beta + Ru , \qquad (5.15)$$

where R $(T \times T)$ is such that $R^{-1}(R\,')^{-1} = V$.

The correct procedure now depends on the null hypothesis. If under H_0 the disturbance covariance matrix is given by (5.14), we have $\text{cov}(\bar{\beta}) = \sigma^2(X\,'V^{-1}X)^{-1}$ and $\text{cov}(\hat{\beta}) = \sigma^2(X'X)^{-1}X\,'VX(X\,'X)^{-1}$, where

$$\text{cov}(\hat{\beta}) \geqq \text{cov}(\bar{\beta}). \qquad (5.16)$$

This implies that the roles of OLS and the comparison estimator are reversed. Apart from this, one proceeds as usual.

White (1980a) suggests a test along these lines, where R is a $T \times T$ diagonal matrix to eliminate heteroskedasticity among disturbances. A useful exposition and critique of this approach can be found in *Breusch* and *Godfrey* (1986).

(iv) The Chow test. We have seen in chapter 4 that the Chow test (4.2) is equivalent to the F-test of H_0: $\Delta\beta = 0$ in the augmented model

$$\begin{bmatrix} y_1 \\ y_2 \end{bmatrix} = \begin{bmatrix} X_1 & 0 \\ X_2 & X_2 \end{bmatrix} \begin{bmatrix} \beta \\ \Delta\beta \end{bmatrix} + \begin{bmatrix} u_1 \\ u_2 \end{bmatrix}$$

On the other hand, we will prove below that any F-test for H_0: $\gamma = 0$ in the general augmented model

$$y = X\beta + Z\gamma + u \qquad (5.17)$$

is equivalent to a data transformation test if the transformation matrix F is related to Z via

$$Z = F\,'F\,X\ .$$

Obviously, this holds for the Chow test if we define $F = [0:I]$, so the Chow test can be viewed as a data transformation test with this transformation matrix.

More generally, any test equivalent to an F-test of H_0: $\gamma = 0$ in the augmented model (5.17), where in addition Z can be expressed as $Z = F\,'FX$, is also equivalent to a data transformation test with transformation matrix F.

The following are examples of data transformation tests based on a comparison of the residual variance:

(v) The Berenblut-Webb test (see p.31). Similar to the Plosser-Schwert-White approach, this test also uses the differencing transformation, i.e.

$$u_t^* = u_t - u_{t-1} \quad (t = 2,...,T)\ .$$

The residual variance of the transformed model should therefore be twice that of the original model if the latter is correct. Accordingly we reject the model whenever the ratio of empirical variances deviate too much from this theoretical value, which is equivalent to rejecting for large values of $|g-2|$.

(vi) King's (1985) $S(\varrho)$ test (see p.32). Here we have

$$u_t^* = u_t - \varrho\ u_{t-1} \quad (t = 2,...,T)$$
$$u_1^* = \sqrt{(1 + \varrho^2)}\ u_1\ ,$$

and the ratio of the residual variance should be $1 + \varrho^2$.

Relationship to IV estimation

Data transformation tests compare the OLS coefficient estimates from the original model (5.1) to the OLS coefficient estimates from the transformed model (5.2). It is however easily seen that the latter estimates amount to a special case of the more general instrumental variables (IV) estimation method. We have explicitly demonstrated this for the differencing test on p. 101 above, and the same holds for *any* estimator obtained by applying OLS to a transformed model of the type (5.2).

An IV estimator for β in the model (5.1) is given by

$$\tilde{\beta} = (Z'X)^{-1}\ Z\,'y \tag{5.18}$$

(compare 4.109), where the columns of the $T \times K$ matrix Z are the instruments. On the other hand, applying OLS to the transformed model (5.2) yields

$$\beta^* = (X\,'F\,'FX)^{-1}\ X\,'F\,'Fy \tag{5.19}$$

(compare 5.4). This shows that β^* is an IV estimator with

$$Z = F\,'FX\ . \tag{5.20}$$

IV estimation requires that the $K \times K$ matrix $Z'X = X'F'FX$ be nonsingular. This is equivalent to our assumption that the transformed regressor matrix $X^* = FX$ has rank K.

Since $\tilde{\beta} = \beta^*$, it does not matter whether we compute a Hausman test via the data transformation approach or via the more general IV approach. In practice, the data transformation approach is often preferable, since a possible rank deficiency in the matrix $(X'F'FX)$ is more easily seen. Also, the instrument matrix in (5.20) is rather artificial and does not allow a natural interpretation.

b) Variable Addition

The second major approach to the construction of indices of model adequacy can be interpreted as the *addition* of selected variables to the equation under test. So far, Ramsey's RESET was the only procedure explicitly based on augmenting the original model and testing the joint significance of the added test variables. We have in addition seen that both version (4.2) and version (4.3) of the Chow test can also be viewed as tests for omitted variables, and similarly the Rainbow test. Next we show that most data transformation tests can likewise be interpreted as RESETs, given a proper choice of the test variables Z in the augmented model $y = X\beta + Z\gamma + u$.

Data transformation tests as RESETs

Consider the augmented regression

$$y = X\beta + F'FX\gamma + u$$
$$= X\beta + Z\gamma + u$$
$$= W\delta + u , \tag{5.21}$$

where $Z = F'FX$, $W = [X:Z]$ and $\delta = [\beta ', \gamma ']'$[1]. For simplicity, let us initially assume that the augmented regressor matrix W has full column rank 2K.

We show next that the classical F-test of H_0: $\gamma = 0$ in the model (5.21) is *algebraically identical* to a version of the data transformation test based on the transformation matrix F.

To see this, reconsider the standard F-test as given in (4.152), with test statistic

$$F = \tilde{\gamma} '[\hat{V} (\tilde{\gamma})]^{-1}\tilde{\gamma}/K , \tag{5.22}$$

where $\tilde{\gamma}$ is a subvector of $\tilde{\delta} = (W'W)^{-1}W'y$ and $\hat{V}(\tilde{\gamma})$ is an estimate for the covariance matrix of $\tilde{\gamma}$.

The OLS estimate for γ in the augmented model (5.21) can be expressed as

$$\tilde{\gamma} = (Z'Q_xZ)^{-1} Z'Q_xy ,$$ (5.23)

where $Q_x = I-X(X'X)^{-1}X'$. Its covariance matrix is therefore

$$V(\tilde{\gamma}) = \sigma^2 (Z'Q_xZ)^{-1} ,$$

and the appropriate estimate for $V(\tilde{\gamma})$ is

$$\hat{V}(\tilde{\gamma}) = \tilde{\sigma}^2 (Z'Q_xZ)^{-1} ,$$ (5.24)

where

$$\tilde{\sigma}^2 = \frac{\tilde{u}'\tilde{u}}{T-2K}$$ (5.25)

and where $\tilde{u}=y-W\tilde{\delta}$ is the OLS residual vector from the augmented regression. This implies that the test statistic (5.22) can also be written as

$$F = y'Q_xZ (Z'Q_xZ)^{-1} Z'Q_xy/K\tilde{\sigma}^2 .$$ (5.26)

On the other hand, we can rewrite the test statistic (5.7) of the data transformation test in a similar way. To this purpose, note that

$$\begin{aligned}
\beta^* &= (Z'X)^{-1}Z'y \\
&= (Z'X)^{-1}Z'(X\hat{\beta}+\hat{u}) \\
&= \hat{\beta} + (Z'X)^{-1}Z'\hat{u} \\
&= \hat{\beta} + (Z'X)^{-1}Z'Q_xy ,
\end{aligned}$$ (5.27)

so the estimator difference is

$$\beta^*-\hat{\beta} = (Z'X)^{-1}Z'Q_xy$$ (5.28)

and

$$\mathrm{cov}(\beta^*-\hat{\beta}) = \sigma^2 (Z'X)^{-1} Z'Q_xZ (X'Z)^{-1} .$$ (5.29)

Together, (5.28) and (5.29) imply that the test statistic (5.7) has the form

$$m = y'Q_xZ (Z'Q_xZ)^{-1}Z'Q_xy/\sigma^2 .$$ (5.30)

In practice σ^2 is unknown and must be replaced by some estimate $\hat{\sigma}^2$, in which case m is asymptotically $\chi^2_{(K)}$, irrespective of the particular estimate. The point is that whenever σ^2 is estimated from (5.25), we obtain the algebraic identity

$$m/K = F ,$$ (5.31)

which means that the asymptotic data transformation test can be made exact and will then have the same rejection region as the standard F-test of H_0: $\gamma=0$.

This important relationship was first discovered by *Breusch* and *Godfrey* (1984), although their proof is incomplete and different from ours. *Davidson* et al. (1986) also prove a similar result.

If the regressor matrix W in the augmented model (5.21) does not have full column rank 2K, the identity (5.31) must be slightly modified.

Let the rank of W be $K + K_1$, where $K_1 < K$. The matrix $Z'Q_xZ$ from the expressions (5.29) and (5.30) has then also rank equal to K_1, and its inverse in (5.30) must be replaced by some g-inverse. The test statistic m for the data transformation test has then K_1 rather than K degrees of freedom, and we have the modified identity $m/K_1 = F$, where the test statistic F is now obtained after dropping any redundant variables in Z.

At the same time this relationship reveals one major advantage of expressing data transformation tests as tests for omitted variables, i.e. rank deficiencies in $cov(\hat{\beta}^* - \hat{\beta})$ are more readily apparent. As an example, consider the differencing test in a model with seasonal dummies. The transformed seasonals are then each linear combinations of the original seasonals, which means that they must be omitted from Z, and that the rank of $cov(\hat{\beta}^* - \hat{\beta})$ must be accordingly reduced.

The F-test interpretation of the differencing test also shows that differencing tests based on pseudo-differences $y_t - \varrho y_{t-1}$ and $x_t - \varrho x_{t-1}$ lead to almost identical results. We have stated this proposition in chapter 4, and can now give an easy proof.

In view of (4.185), a typical row of Z is after pseudo-differencing given by

$$z_t = (1 + \varrho^2) x_t - \varrho x_{t+1} - \varrho x_{t-1} . \qquad (5.32)$$

This holds for all rows if we drop the first and the last observation in the augmented model (5.21). Subtracting $(1 + \varrho^2)x_t$ from z_t then does not change the value of the F-statistic (5.22). Neither does the multiplication of this modified Z matrix by ϱ^{-1}, which however produces identical test variables *regardless of* ϱ. This shows that any numerical difference between the respective test statistics can derive only from the first and last observation in the sample, and will therefore tend to zero in probability regardless of whether H_0 is true or not. Contrary to *Plosser* et al. (1982), the tests are thus asymptotically equivalent. This fact was first noted by *Breusch* and *Pagan* (1984).

Summing up, we have now altogether *three* seemingly different tests which have turned out algebraically identical. The first is the standard F-test for H_0: $\gamma = 0$ in the augmented regression $y = X\beta + F'FX\gamma + u$. The second is the Hausman test based on a comparison of the OLS estimates for β in the original and in the augmented model. We have shown in chapter 4 that this is identical to the F-test if the number of test variables does not exeed the number of regressors in the original model - a condition that is obviously satisfied here. The third test finally is the Hausman test based on the transformed data, which again amounts to the simple F-test of H_0: $\gamma = 0$.

Hausman tests as RESETs

Similar to data transformation tests, *any* Hausman test where OLS is compared to IV can easily be established as a RESET. We have directly shown this for the Wu test in section 4.c, and the same argument covers the general case as well.

To see this, let $\tilde{\beta} = (Z\,'X)^{-1} Z\,'y$ be the relevant IV estimator, with T by K instrument matrix Z. If $W = [X:Z]$ has rank 2K, the resulting Hausman test is algebraically equivalent to the F-test of H_0: $\gamma = 0$ in the model $y = X\beta + Z\gamma + u$, by simply applying the argument following (5.21). If W has less than full column rank, we again drop any redundant columns and reduce the degrees of freedom accordingly.

Since almost all econometric estimators can be exhibited as IV estimators of some sort, this shows that tests for omitted variables are by far the most important diagnostic technique in econometrics, although, due to various guises, they may not immediately be recognized as such.

Power comparisons

The RESET interpretation of misspecification tests is useful in various ways. For instance, it helps evaluating the relative power of competing tests, by providing a unified framework for discussing the non-null distribution of the test statistics.

Consider the standard model $y = X\beta + u$, where $E(u) = \xi \neq 0$ under the alternative. Let two competing tests be based on the augmented regressions

$$y = X\beta + Z_1\gamma_1 + u \tag{5.33}$$

and

$$y = X\beta + Z_2\gamma_2 + u, \tag{5.34}$$

respectively, where Z_1 is T by K_1 and Z_2 is T by K_2. For simplicity, let both augmented regressions have full column rank. Which test has higher power?

Following *Thursby* and *Schmidt* (1977), we briefly examine this issue for the case where Z_1 and Z_2 are nonstochastic and where the disturbances are normal and homoskedastic *also under the alternative*. The test statistics for the respective RESETs then have a doubly noncentral F distribution with K_i and $T-K-K_i$ degrees of freedom. The numerator noncentrality parameters are

$$\lambda_i = \xi' Q_x Z_i (Z_i' Q_x Z_i)^{-1} Z_i' Q_x \xi / \sigma^2 \quad (i = 1,2) \tag{5.35}$$

and the denominator noncentrality parameters are

$$\delta_i = \xi' [Q_x - Q_x Z_i (Z_i' Q_x Z_i)^{-1} Z_i' Q_x] \xi / \sigma^2 \quad (i = 1,2), \tag{5.36}$$

respectively (see Thursby and Schmidt, 1977, p.635). The denominator noncentrality parameters would equal zero if and only if $\xi = Z_1\gamma_i$ for some γ_i, that is, if the nonzero disturbance mean were generated by omission of certain variables in Z_1.

Given the degrees of freedom, the power of an F-test is an increasing function of the numerator noncentrality parameter, and a decreasing function of the denominator noncentrality parameter. A sufficient condition for the test based on (5.33) to have higher power than the test based on (5.34) is therefore

$$K_1 = K_2, \lambda_1 > \lambda_2, \delta_1 < \delta_2. \tag{5.37}$$

Failing (5.37), the determination of the ranking among the tests is a computer job, and nothing much can be said analytically. If both denominator noncentrality parameters are zero, on can use *Han*'s (1985) algorithm to compute the respective rejection probabilities for any given ξ. If denominator noncentrality parameters are nonzero, the noncentral F distribution becomes very complicated, and one has to rely on approximations such as given in Thursby and Schmidt (1975, p.636).

It is obvious from (5.35) that $\lambda_i = 0$, i.e. that RESET has no power, if

$$\xi \, 'Q_x Z_1 = 0 \, ,$$

confirming our earlier result from p.100. Moreover, the denominator noncentrality parameter can still be positive, in which case the power of the test could even shrink below its size.

It is of course unreasonable to assume that the disturbances will remain normal and homoskedastic under misspecification, so the above analysis is of mainly academic interest.

Asymptotic independence

A major advantage of discussing misspecification tests in an omitted variable framework is that it is then easy to determine whether or not two given tests are independent, at least asymptotically. This in turn is important to know whenever several tests are applied to identical data sets at a time, in order to compute the overall rejection probability of the joint procedure. Such issues will be considered in more detail in section 6.b below.

For the moment, consider the F-tests of H_0: $\gamma_i = 0$ in the equations (5.33) and (5.34), respectively. When will these tests be independent?

We compute the test statistics from the formula (5.26) and consider the numerators first. Obviously, these are independent if and only if

$$Z_1 \, 'Q_x Z_2 = 0 \, . \tag{5.38}$$

Since

$$Z_1 \, 'Q_x Z_2 = Z_1 \, (I - X(X \, 'X)^{-1} X \, ') \, Z_2$$
$$= Z_1 \, 'Z_2 - Z_1 \, ' \, X(X \, 'X)^{-1} X \, 'Z_2 \, ,$$

a sufficient condition for (5.38) is that the columns of Z_1 and Z_2, and the columns of either Z_1 and X or Z_2 and X are pairwise orthogonal.

Given (5.38), the tests would be exactly independent if $\tilde{\sigma}^2$ from the denominator of (5.26) were a fixed constant. Since $\tilde{\sigma}_i^2 \xrightarrow{p} \sigma^2$ under H_0, where

$$\cdot \quad \tilde{\sigma}_i^2 = (y\text{-}X\hat{\beta}\text{-}Z_i\hat{\gamma}_i)\,'(y\text{-}X\hat{\beta}\text{-}Z_i\hat{\gamma}_i)\,/(T\text{-}K\text{-}K_i), \qquad\qquad (5.39)$$

the stochastic nature of the denominator does however not matter asymptotically, and the tests will be asymptotically independent under H_0, provided

$$Z_1\,'Q_xZ_2\,/\,T \to 0\,. \qquad\qquad (5.40)$$

The asymptotic independence does not require normality of the disturbances. It follows from the joint normal limiting distributions of $\sqrt{T}\hat{\gamma}_1$ and $\sqrt{T}\hat{\gamma}_2$, which have limiting covariance matrix

$$\lim_{T\to\infty} (Z_1\,'Q_xZ_1/T)^{-1}\,(Z_1Q_xZ_2/T)\,(Z_2Q_xZ_2/T)^{-1}\,.$$

As always, we tacitly assume here that all required limits exist and are nonsingular if necessary.

6. DIAGNOSTIC CHECKING IN PRACTICE

This chapter is devoted to some real life applications of the methods discussed so far, plus an examination of the statistical problems of applying several tests to identical data sets at a time (see also *Krämer, Sonnberger, Maurer* and *Havlik*, 1985).

a) Empirical Results

Section a) confronts several regression equations from the empirical literature with the diagnostic checks and tests from chapters three and four. We focus on tests for the regressor specification, since it is here that the linear regression model stands or falls. Tests for the correct specification of the second moments of the disturbances are included only when they also serve to detect inadequacies in the deterministic part of the equation. We in addition confine ourselves to static models so that we can apply a maximum number of tests.

The only criterion for including a particular study was data availability. All empirical results are from well established journals and easily accessible. The data are those used by the original authors and are reproduced in the appendix.

We test only one equation from each empirical study - the first one that meets the criteria (i) that there are no lagged endogenous variables in the equation and (ii) that the equation is linear in the parameters - in order to avoid any intentional malice on our part. All empirical work the data of which we could get hold on and meeting these restrictions was included. In studies that mainly vary previous work, we consider the first equation that differs from the previous investigation.

The empirical results scrutinized below are certainly no random sample from the literature. The single equation models that we are testing might be accused of being oversimplified and thus particularly prone to fail our specification tests. A different and possibly offsetting bias might be introduced by the type of data used for many studies, which mostly concern monetary phenomena. To the extend that monetary data are more accurate than other data used in applied regression analysis, one should expect such studies to fare better than average. Also, the fact that the data were made available for outside checking, either by the authors themselves or by the journals where the papers have appeared, is by itself an indicator of quality.

Demand for money

We first test a model suggested by *Klein* (1977) for the demand for money in the United States. The data are from *Allen* (1982), where Klein's model is slightly revised, and were given to us by the Journal of Money, Credit and Banking.

Klein and Allen investigate the effect of uncertainty regarding future inflation rates on long run money demand. The argument is that increased inflation uncertainty lowers the stream of monetary services yielded by a given level of real cash balances. While economic theory is ambiguous about the resulting effect on the demand for money, Klein's empirical results show that the demand for money is always increased by an increase in price uncertainty.

We focus here on equation (10) from Klein (1977, p. 703, table 1). This is the first empirical equation in the paper where price uncertainty enters as an explanatory variable and reads (Klein's notation)

$$\log M = a_0 + a_1 \log y_p + a_2 r_s + a_3 r_L + a_4 r_M + a_5 \log S(\dot{p}/p), \tag{6.1}$$

where M is the quantity of money (M_2), y_p is real permanent income, r_s is a short term interest rate, and r_M is the rate of return on money.

Of primary interest here is the variable $S(\dot{p}/p)$, an operational measure of the *variability* of the rate of price changes. It is defined as a five-term moving standard deviation from the 10-term moving mean of the annual rate of change of prices and is plotted in Klein (1977, p. 701, fig. 2). Its sign in Klein's regressions is always significantly positive.

Allen (1982) argues that the variable r_M in (6.1) is really endogenous in a larger simultaneous equation fromework, and that Klein's empirical results therefore suffer from simultaneous equation bias. He shows that if r_M is omitted from equation (6.1), the positive relationship between $\log S(\dot{p}/p)$ and $\log M$ turns insignificant. In addition, Allen corrects for autocorrelation by a Cochrane-Orcutt transformation:

$$
\begin{aligned}
y_t^* &= y_t - \varrho y_{t-1} \\
x_{tk}^* &= x_{tk} - \varrho x_{t-1,k} \quad (t=2,\dots,T; \ k=1,\dots,K)
\end{aligned}
\tag{6.2}
$$

where $\varrho = .989$. He does not specify whether or not the first transformed data point is omitted. We computed it as

$$
\begin{aligned}
y_1^* &= (1-\varrho^2)^{1/2} y_1 \\
x_{1k}^* &= (1-\varrho^2)^{1/2} x_{1k} \quad (k=1,\dots,K) .
\end{aligned}
\tag{6.3}
$$

Below we focus on Allen's equation (1.4) (p.527, table 1).

Table (6.1) reproduces the respective regression results, both as reported by the original authors and as computed by ourselves, with t-values in parentheses below the estimated regression coefficients. The data are annual, from 1880 to 1972, and are reproduced in table (A.1) in the appendix.

TABLE 6.1: MONEY DEMAND REGRESSION

	explanatory variables						
	const	$\log Y_p$	r_S	r_L	r_M	$\log S(\dot{p}/p)$	d
Klein	-14.01	1.380	-.268	-.071	.302	.050	.98
	(69.55)	(41.44)	(11.37)	(5.86)	(11.09)	(4.48)	
OLS	-13.74	1.347	-.282	-.077	.320	.059	1.07
	(69.36)	(40.45)	(11.89)	(6.17)	(11.65)	(5.11)	
Allen	-13.61	1.25	-.005	-.059		.014	1.61
	(18.32)	(11.25)	(1.09)	(3.74)		(.95)	
CO	-13.62	1.25	-.005	-.059		.014	1.61
	(18.39)	(11.28)	(1.09)	(3.74)		(.95)	
OLS	-15.49	1.62	-.019	-.084		.084	.18
	(75.09)	(42.88)	(1.65)	(4.23)		(4.65)	

The estimates which we obtained ourselves are always in the rows denoted "OLS" (when we did an Ordinary Least Squares regression) or "CO" (when we did a Cochrane-Orcutt transformation.

There are only minor discrepancies between our regression results and Allen's, which we attribute to different estimation algorithms (the one that we used - the IAS-SYSTEM - is described in more detail in the appendix). The differences between our results and Klein's are most probably due to data revisions subsequent to Klein's study.

Table (6.2) summarizes the results of subjecting equation (6.1) to our various tests. Part b refers to Allen's version of equation (6.1), where OLS and the tests are applied to the data after the Cochrane-Orcutt transformation. In part c, both OLS and the tests are applied to the untransformed data.

The first column in the table specifies the test, the second gives the value of the test statistic, the third the (asymptotic) null distribution, the fourth the prob-value of the test-statistic, and the last column reports whether or not the null hypothesis (i.e. that the assumptions of the standard linear regression model hold) is rejected at the 5% significance level. The "-" sign means pass and "x" signifies failure. Numbers in parentheses next to the name of the test indicate for which period a rejection could be obtained (outlier test), where the sample has been split (Chow test), or where the CUSUM statistic first crossed a critical 5% line. "n.c." means "not computable". This can occur due to software restrictions, or because matrices are singular.

Table 6.2: TEST RESULTS FOR MONEY DEMAND REGRESSIONS

Test	value of test stat.	distr. under H_0	prob-value	rej. at 5 %
a) Klein (OLS)				
Chow (1950)	7.133	$F_{(6,81)}$	.00 %	x
Breusch-Pagan	14.457	$\chi^2_{(5)}$	1.30 %	x
Breusch-Pagan (s)	15.504	$\chi^2_{(5)}$	.84 %	x
CUSUM				-
CUSUM (s)				-
CUSUM of sq. (1909)				x
CUSUM of sq. (s)	2.022			-
Fluct.	8.582	1.650		x
RESET (a)	9.397	$F_{(2,85)}$	.02 %	x
RESET (b)	18.587	$F_{(5,82)}$	.00 %	x
RESET (c)	12.853	$F_{(2.85)}$	.00 %	x
Harvey-Collier (1)	-2.564	$t_{(86)}$	1.21 %	x
Harvey-Collier (2)				n.c.
Harvey-Collier (3)	2.121	$t_{(86)}$	3.68 %	x
Harvey-Collier (4)	3.254	$t_{(86)}$	.16 %	x
Harvey-Collier (5)	-1.223	$t_{(86)}$	22.48 %	-
Godfrey-Wickens Log (1,5)	9.035	$\chi^2_{(1)}$	.26 %	x
Rainbow	2.642	$F_{(47,40)}$	.11 %	x
Outlier (1893)	3.391			-
Hausman	34.417	$\chi^2_{(5)}$	.00 %	x
Diff.	36.106	$\chi^2_{(5)}$	.00 %	x
White				n.c.
IMT	45.487	$\chi^2_{(28)}$	1.97 %	x
b) Allen (CO)				
Chow (1960)	1.518	$F_{(5,83)}$	19.31 %	-
Breusch-Pagan	13.171	$\chi^2_{(4)}$	1.05 %	x
Breusch-Pagan (s)	11.273	$\chi^2_{(4)}$	2.37 %	x
CUSUM				-
CUSUM (s)				-
CUSUM of sq.				-
CUSUM of sq. (s)	.885			-
Fluct.	2.562	1.620		x

Table 6.2 (continued)

RESET (a)	6.614	F (2,86)	.21 %	x
RESET (b)	5.571	F (4,84)	.05 %	x
RESET (c)	.035	F (2,86)	96.59 %	-
Harvey-Collier (1)	-.815	t (87)	41.47 %	-
Harvey-Collier (2)	-.020	t (87)	98.44 %	-
Harvey-Collier (3)	-.349	t (87)	72.82 %	-
Harvey-Collier (4)	3.032	t (87)	.32 %	x
Godfrey-Wickens Log (1,4)	15.930	χ^2 (1)	.01 %	x
Rainbow	1.210	F (47,41)	26.79 %	-
Outlier (1880)	3.646			x
Hausman	13.423	χ^2 (4)	.94 %	x
Diff.				n.c.
White	2.118	χ^2 (15)	100 %	-
IMT	31.230	χ^2 (21)	6.99 %	-
		c) Allen (OLS)		
Chow (1913)	58.088	F (5,83)	.00 %	x
Breusch-Pagan	11.681	χ^2 (4)	1.99 %	x
Breusch-Pagan (s)	16.635	χ^2 (4)	.23 %	x
CUSUM (1953)				x
CUSUM (s)				x
CUSUM of sq. (1903)				x
CUSUM of sq. (s)	2.571			?
Fluct.	9.189	1.620		x
RESET (a)	92.702	F (2,86)	.00 %	x
RESET (b)	68.175	F (4,84)	.00 %	x
RESET (c)	2.949	F (2,86)	5.77 %	-
Harvey-Collier (1)	-5.562	t (87)	.00 %	x
Harvey-Collier (2)	1.441	t (87)	15.31 %	-
Harvey-Collier (3)	.628	t (87)	53.19 %	-
Harvey-Collier (4)	-.921	t (87)	35.95 %	-
Godfrey-Wickens Log (1,4)	54.486	χ^2 (1)	.00 %	x
Rainbow	1.141	F (47,41)	33.54 %	-
Outlier (1880)	2.565			-
Hausman	2.838	χ^2 (4)	58.52 %	-
Diff.	4.144	χ^2 (4)	38.69 %	-
White	54.579	χ^2 (15)	.00 %	x
IMT	32.401	χ^2 (21)	5.33 %	-

We begin by testing for structural change. The Chow test is done as described in section 4.a, with test statistic (4.2). The sample was split where the Quandt-Ratio (4.7) was smallest (which of course inflates the size of the test well above the nominal 5%). The Breusch-Pagan test, which was discussed in section 3.b under the heading of heteroskedasticity, is also included here, since it is also sensitive to stochastically varying parameters (see 3.82). The test statistics are either (3.75), or Koenker's (1981) studentized version thereof. The latter test is distinguished by an "s" (for "studentized").

Similarly, we distinguish between the standard and the studentized version of the CUSUM of squares test.

The CUSUM test is always the one where the disturbance variance is estimated by (4.73). "Fluct." stands for the Fluctuation test.

Next we consider tests for functional form. RESET(a) means that second and third powers of the fitted dependent variable are used to form the Z matrix in the auxiliary regression (4.82). In RESET(b), we use the second powers of the independent variables, and second and third powers of the first principal component of the regressor matrix X are used in RESET(c). The various versions of the Harvey-Collier ψ-test (see 4.84) are distinguished according to the number of the independent variable under test. Harvey-Collier (3) for instance means that the sample was reordered according to increasing values of the third right-hand side variable (excluding the constant) in equation (6.1).

The Godfrey-Wickens and Outlier tests are performed as described in section 4.b. The test statistic for the Rainbow test is given in (4.83). We always used the 50% data points with smallest leverage for the auxiliary regression.

The remaining procedures test for general misspecification. We always used lagged values of the independent variables as instruments for the Hausman test. "Diff." stands for the differencing test, "White" is White's (1980) test for heteroskedasticity from section 4.b, which is also consistent against misspecification alternatives, and "IMT" stands for the Information Matrix Test.

Table (6.2) illuminates two things. First, all versions of equation (6.1), whether estimated from the transformed or from the untransformed data, are rejected more often than can be expected by chance. We will make this more precise below. Second, the Cochrane-Orcutt transformation reduces the failure rate - a feature that will reappear in other models as well, and which points to a sensibility of the tests to autocorrelation among disturbances.

We will return to these topics after we have seen more test results below.

Currency substitution

Next we examine two related studies by *Miles* (1978) and *Bordo and Choudhri* (1982) which focus on the effect of currency substitution on the demand for money. Currency substitution means that residents of one country substitute foreign and domestic currencies in their money holdings in response to changes in relative interest rates and related exogenous variables. If currency substitution were empirically important, it would transmit monetary disturbances from one country to another and would thus seriously undermine the ability of flexible exchange rates to provide monetary independence.

Miles (1978) argues that this is indeed the case for Canada. His empirical model is (Miles' notation; see eq. 12, p. 434)

$$\log(\frac{C \, \$}{US \, \$}) = \beta_1 + \beta_2 \log(\frac{1+i_{US}}{1+i_C}) , \tag{6.4}$$

where

C $:	Canadian holdings of Canadian dollar balances,
US $:	Canadian holdings of U.S. dollar balances,
i_{US}:	Yield on U.S. Treasury bills, and
i_C:	Yield on Canadian Treasury bills.

Miles suggests that when U.S. money becomes more costly (i.e. when the exogenous variable in (6.4) increases), Canadians will substitute Canadian currency for U.S. currency in their money holdings. This is supported by his empirical regression, which produces a large positive value for the coefficient β_2.

Bordo and *Choudhri* (1982) question the significance of this relationship. They argue that the estimate for β_2 becomes insignificant when additional regressors are introduced in (6.4). Their model is (p. 54, eq. 1, table 2)

$$\log(\frac{C \, \$}{US \, \$}) = \beta_1 + \beta_2(i_{US} - i_C) + \beta_3 i_C + \beta_4 \log y , \tag{6.5}$$

where the inclusion of the additional regressors i_C and y (= Canadian real gross national product) is justified by theoretical arguments which we will not repeat here. The use of $i_{US} - i_C$ rather than $\log((1+i_{US})/(1+i_C))$ in (6.5) does not make much difference, since both quantities are roughly proportional in the range of the observed values.

Table (6.3) gives the regression results. Data are quarterly, covering the period 1960.IV-1975.IV, and are reproduced in table A.2 in the appendix. Again, this is to invite readers to check our results.

Both Miles and Bordo and Choudhri applied a Cochrane-Orcutt transformation prior to OLS, with ϱ-values as indicated in the table. Similar to table 6.1, t-values are given in parentheses below the coefficient estimates.

This time, our regression results do not agree that well with the original authors. This may be due to subsequent data revisions as far as Miles is concerned, but cannot be so explained for the Bordo-Choudhri equation, since we used their data. We omitted the first observation for our CO-regression. Otherwise, the discrepancy would have been even more pronounced.

Table (6.4) reports the test results. The remarks concerning table (6.2) apply here as well. In particular, certain tests such as RESET and WHITE were not always applicable.

The equations under test in table (6.4) fare much better than those from table (6.2). Otherwise, the earlier picture is confirmed: both equations do worse without a CO-transformation. Again, this provides some evidence that our tests often suggest a structural change or incorrect functional form when the real culprit is "only" autocorrelation.

Table 6.3: CURRENCY SUBSTITUTION REGRESSIONS

	explanatory variables						
	const.	$\log(\frac{1+i_{us}}{1+i_c})$	$i_{us}-i_c$	i_c	y	d	ρ
Miles	2.56 (18.00)	5.43 (2.59)				1.44	.88
CO	2.60 (20.10)	4.82 (2.15)				1.57	.88
OLS	2.56 (73.25)	5.98 (1.74)				.26	
Bordo-Choudhri	-5.72 (1.13)		-1.39 (.59)	-1.40 (4.35)	.78 (1.74	1.70	.85
CO	-5.21 (.98)		-1.51 (.62)	-10.90 (3.99)	.74 (1.56)	1.78	.85
OLS	-6.24 (3.22)		-9.39 (2.86)	-18.66 (7.36)	.86 (4.78)	.51	

Table 6.4: TEST RESULTS FOR CURRENCY SUBSTITUTION REGRESSIONS

Test	value of test stat.	distr. under H_o	prob-value	rej. at 5 %
		a) Miles (CO)		
Chow (1974.II)	1.330	F (2,57)	27.26 %	-
Breusch-Pagan	.043	χ^2 (1)	83.62 %	-
Breusch-Pagan (s)	.074	χ^2 (1)	78.63 %	-
CUSUM				-
CUSUM (s)				-
CUSUM of sq.				-
CUSUM of sq. (s)	1.063			-
Fluct.	.795	1.480		-
RESET (a)	1.177	F (2,57)	31.55 %	-
RESET (b)	1.582	F (1,58)	21.35 %	-
Harvey-Collier (1)	-2.662	t (58)	1.00 %	x
Godfrey-Wickens Log (1)	.174	χ^2 (1)	67.65 %	-
Rainbow	.784	F (31.28)	74.54 %	-
Outlier (1969.II)	2.439			-
Hausman	.969	χ^2 (1)	32.48 %	-
Diff.	1.192	χ^2 (1)	27.50 %	-
White				n.c.
IMT	4.975	χ^2 (6)	54.70 %	-
		b) Miles (OLS)		
Chow (1964.II)	4.913	F (2,57)	1.08 %	x
Breusch-Pagan	.246	χ^2 (1)	61.98 %	-
Breusch-Pagan (s)	.285	χ^2 (1)	59.34 %	-
CUSUM (1969.III)				x
CUSUM (s)				x
CUSUM of sq. (1965.III)				x
CUSUM of sq. (s)	2.012			x
Fluct.	3.108	1.480		x
RESET (a)	1.089	F (2,57)	34.35 %	-
RESET (b)	2.040	F (1,58)	15.68 %	-
RESET (c)				n.c.
Harvey-Collier (1)	.932	t (58)	35.50 %	-
Godfrey-Wickens Log (1)	.096	χ^2 (1)	75.70 %	-
Rainbow	1.643	F (31,28)	9.39 %	-
Outlier (1972.III)	2.386			-
Hausman	.070	χ^2 (1)	79.13 %	-
Diff.	.031	χ^2 (1)	85.98 %	-
White	3.200	χ^2 (3)	36.18 %	-
IMT	10.881	χ^2 (6)	9.21 %	-

Table 6.4 (continued)

c) Bordo-Choudhri (CO)

Chow (1961.IV)	2.383	$F_{(3,53)}$	7.97 %	–
Breusch-Pagan	.404	$\chi^2_{(3)}$	93.94 %	–
Breusch-Pagan (s)	.624	$\chi^2_{(3)}$	89.10 %	–
CUSUM				–
CUSUM (s)				–
CUSUM of sq.				–
CUSUM sq. (s)	.649			–
Fluct.	2.081	1.590		x
RESET (a)	.689	$F_{(2,54)}$	50.66 %	–
RESET (b)	1.831	$F_{(3,53)}$	15.27 %	–
RESET (c)	2.332	$F_{(2,54)}$	10.68 %	–
Harvey-Collier (1)	-1.891	$t_{(55)}$	6.39 %	–
Harvey-Collier (2)	.444	$t_{(55)}$	65.90 %	–
Harvey-Collier (3)	1.843	$t_{(55)}$	7.08 %	–
Godfrey-Wickens Log (3)	.205	$\chi^2_{(1)}$	65.06 %	–
Rainbow	1.100	$F_{(30,26)}$	40.52 %	–
Outlier (1961.I)	2.280			–
Hausman	16.122	$\chi^2_{(3)}$	.11 %	x
Diff.	5.359	$\chi^2_{(3)}$	14.73 %	–
White	12.005	$\chi^2_{(10)}$	28.47 %	–
IMT	12.653	$\chi^2_{(15)}$	62.90 %	–

d) Bordo-Choudhri (OLS)

Chow (1970.II)	14.868	$F_{(4,53)}$	.00 %	x
Breusch-Pagan	13.239	$\chi^2_{(3)}$	.41 %	x
Breusch-Pagan (s)	10.673	$\chi^2_{(3)}$	1.36 %	x
CUSUM				–
CUSUM (s)				–
CUSUM of sq. (1966.II)				x
CUSUM of sq. (s)	2.072			x
Fluct.	5.037	1.590		x
RESET (a)	.781	$F_{(2,55)}$	46.31 %	–
RESET (b)	7.221	$F_{(3,54)}$	.04 %	x
RESET (c)	9.293	$F_{(2,55)}$	.03 %	x
Harvey-Collier (1)	.828	$t_{(56)}$	41.14 %	–
Harvey-Collier (2)	.294	$t_{(56)}$	77.01 %	–
Harvey-Collier (3)	2.662	$t_{(56)}$	1.01 %	x
Godfrey-Wickens Log (3)	1.249	$\chi^2_{(1)}$	26.37 %	–
Rainbow	3.623	$F_{(31,26)}$	.06 %	x
Outlier (1971.I)	3.156			–
Hausman	16.662	$\chi^2_{(3)}$	.08 %	x
Diff.	4.477	$\chi^2_{(3)}$	21.44 %	–
White	13.021	$\chi^2_{(10)}$	22.25 %	–
IMT				n.c.

Bond yield

Interest rates for different securities do not move parallel in many economies. In the United States there have been several attempts to explain the changing yield differential between U.S. government bonds and corporate bonds. Below we examine the work by *Cook* and *Hendershott* (1978) and *Yawitz* and *Marshall* (1981).

Cook and Hendershott (1978, p. 1183, eq. 1, table 2) estimate the following regression (among others):

$$RAa - RUS = \beta_1 + \beta_2 MOOD + \beta_3 EPI + \beta_4 EXP + \beta_5 RUS, \qquad (6.6)$$

where

RAa:	interest rate on government bonds,
RUS:	interest rate on corporate bonds,
MOOD:	measure of consumer sentiment,
EPI:	index for employment pressure, and
EXP:	interest rate expectations.

The rationale behind this model is as follows: The yield spread RAa-RUS is sensitive to the risk of default for corporate bonds, which in turn varies with the business cycle, motivating the MOOD and EPI explanatory variables. The RUS series is included as a separate regressor to catch the effect on the spread when yields rise with risk held constant. The EXP variable finally measures the effect of the callability of corporate bonds on the yield spread: when future interest rates are high, investors expect firms to exercise their call option to refinance their dept at more favorable terms. As a result, the bondholder will require compensation in the form of a higher coupon on the callable bond for equal initial dollar outlays.

Yawitz and *Marshall* (1981) propose to measure the effect of callability differently, replacing the EXP variable by two artificial time series y and K, respectively, which basically depend on the initial price difference between two bonds. Their equation is (see Yawitz and Marshall, 1981, p. 69, table 3, eq. Y-M1)

$$RAa - RUS = \beta_1 + \beta_2 MOOD + \beta_3 y + \beta_4 K \qquad (6.7)$$

Table (6.5) summarizes the results of the various regressions, both as obtained by the original authors and by ourselves. Data are quarterly, from 1961.I to 1975.IV, and are reproduced in table (A.3) in the appendix. Again, a Cochrane-Orcutt transformation was applied in both regressions prior to OLS, with ϱ-values as shown in the table.

The regression results conform to a priori expectations. In particular, the coefficients of the cyclical default risk variables, MOOD and EPI, and the call risk variables, EXP, all have the correct sign in the regression (6.7). Again there are differences between our results and the literature, which we contribute to different regression algorithms (Yawitz-Marshall) or an incomplete match of the data (Cook-Hendershott).

Table (6.6) summarizes the results of subjecting equations (6.6) and (6.7) to our tests. Similar to tables (6.2) and (6.4), we test the equation also as estimated by OLS, even when the original authors apply a Cochrane-Orcutt transformation.

Table 6.5: BOND YIELD REGRESSIONS

	const.	MOOD	EPI	EXP	RUS	y	K	d	ρ
			explanatory variables						
Cook-Hendershott	11.36	-.0116	-.0977	-.6698	.0205			2.02	.74
	(2.23)	(2.40)	(1.97)	(3.32)	(.42)				
CO	11.49	-.0120	-.0982	-.6818	.0168			2.01	.74
	(2.25)	(2.48	(1.98)	(3.37)	(.34)				
OLS	.784	-.0014	-.0081	-.1749	-.1370			.84	
	(.17)	(.31)	(.18)	(.73)	(3.87)				
Yawitz-Marshall	-5.42	-.012				4.077	176.4		.267
	(2.28)	(2.91)				(5.39)	(3.40)		
CO	-5.46	-.012				4.799	177.4	1.86	.267
	(2.31)	(2.91)				(5.44)	(3.43)		
OLS	-5.07	-.013				4.900	170.9	1.44	
	(2.44)	(3.62)				(6.63)	(3.81)		

Table 6.6: TEST RESULTS FOR BOND YIELD REGRESSIONS

Test	value of test stat.	distr. under H_o	prob-value	rej. at 5 %
		a) Cook-Hendershott (CO)		
Chow (1962.II)	2.047	F (5,50)	8.79 %	-
Breusch-Pagan	1.285	χ^2 (4)	86.39 %	-
Breusch-Pagan (s)	1.483	χ^2 (4)	82.97 %	-
CUSUM (1972.III)				x
CUSUM (s) (1972.III)				x
CUSUM of sq.				-
CUSUM of sq. (s)	.775			-
Fluct.	6.596	1.620		x
RESET (a)	1.016	F (2,53)	36.91 %	-
RESET (b)	1.092	F (4,51)	37.03 %	-
RESET (c)	1.855	F (2,53)	16.65 %	-
Harvey-Collier (1)	-2.561	t (54)	1.33 %	x
Harvey-Collier (2)	-.630	t (54)	53.16 %	-
Harvey-Collier (3)	-.663	t (54)	51.01 %	-
Harvey-Collier (4)	-1.337	t (54)	18.67 %	-
Godfrey-Wickens Lin				n.c.
Rainbow	1.990	F (30,25)	4.14 %	x
Outlier (1973.I)	3.119			-
Hausman	8.655	χ^2 (4)	7.03 %	-
Diff.				n.c.
White	14.145	χ^2 (15)	51.46 %	-
IMT	19.739	χ^2 (21)	53.78 %	-
		b) Cook-Hendershott (OLS)		
Chow (1971.I)	12.743	F (5,50)	.00 %	x
Breusch-Pagan	2.019	χ^2 (4)	73.22 %	-
Breusch-Pagan (s)	3.480	χ^2 (4)	48.10 %	-
CUSUM				-
CUSUM (s)				-
CUSUM of sq. (1967.I)				x
CUSUM of sq. (s)	2.385			x
Fluct.	14.445	1.620		x
RESET (a)	5.180	F (2,53)	.88 %	x
RESET (b)	4.523	F (4,51)	.33 %	x
RESET (c)	6.756	F (2,53)	.24 %	x
Harvey-Collier (1)	-2.370	t (54)	2.14 %	x
Harvey-Collier (2)	-.155	t (54)	87.73 %	-
Harvey-Collier (3)	.507	t (54)	61.39 %	-
Harvey-Collier (4)	-1.576	t (54)	12.08 %	-
Godfrey-Wickens Lin				n.c.
Rainbow	1.708	F (30,25)	8.78 %	-
Outlier (1966.IV)	2.289			-
Hausman	13.738	χ^2 (4)	.82 %	x
Diff.	8.425	χ^2 (4)	7.72 %	-
White	28.293	χ^2 (15)	1.98 %	x
IMT				n.c.

Table 6.6 (continued)

c) Yawitz-Marshall (CO)

Chow (1966.I)	2.538	F (4,52)	5.08 %	-
Breusch-Pagan	6.037	χ^2 (3)	10.98 %	-
Breusch-Pagan (s)	5.130	χ^2 (3)	16.25 %	-
CUSUM				-
CUSUM (s)				-
CUSUM of sq.				-
CUSUM of sq. (s)	1.025			-
Fluct.	15.959	1.590		x
RESET (a)	1.361	F (2,54)	26.51 %	-
RESET (b)	.928	F (3,53)	43.38 %	-
RESET (c)	1.488	F (2,54)	23.50 %	-
Harvey-Collier (1)	-.348	t (55)	72.90 %	-
Harvey-Collier (2)	1.612	t (55)	11.28 %	-
Harvey-Collier (3)	1.552	t (55)	12.64 %	-
Godfrey-Wickens Lin				n.c.
Rainbow	1.184	F (30,26)	33.31 %	-
Outlier (1961.II)	3.499			-
Hausman	10.301	χ^2 (3)	1.62 %	x
Diff.	2.117	χ^2 (3)	54.84 %	-
White	9.789	χ^2 (10)	45.92 %	-
IMT	74.271	χ^2 (15)	.00 %	x

d) Yawitz-Marshall (OLS)

Chow (1966.I)	2.852	F (4,52)	3.27 %	x
Breusch-Pagan	5.814	χ^2 (3)	12.10 %	-
Breusch-Pagan	6.088	χ^2 (3)	10.74 %	-
CUSUM				-
CUSUM (s)				-
CUSUM of sq.				-
CUSUM of sq. (s)	1.239			-
Fluct.	22.573	1.590		x
RESET (a)	2.199	F (2,54)	12.07 %	-
RESET (b)	4.059	F (3,53)	1.14 %	x
RESET (c)	1.995	F (2,54)	14.59 %	-
Harvey-Collier (1)	.241	t (55)	81.05 %	-
Harvey-Collier (2)	2.094	t (55)	4.09 %	x
Harvey-Collier (3)	.061	t (55)	95.19 %	-
Godfrey-Wickens Lin				n.c.
Rainbow	1.289	F (30,26)	25.73 %	-
Outlier (1961.II)	3.260			-
Hausman	4.761	χ^2 (3)	19.02 %	-
Diff.	5.342	χ^2 (3)	14.84 %	-
White	13.651	χ^2 (10)	18.95 %	-
IMT				n.c.

Growth of money supply

There has been some debate in the United States on whether or not the Federal Reserve System in the 1970s used the money supply as its predominant intermediate target of policy. Below we examine a study by *Hetzel* (1981) which concludes that it did not.

Hetzel argues as follows: If the Federal Open Market Committee had indeed viewed the money supply as its primary intermediate target of policy, it would have provided for operating procedures that would have ensured control of the money supply. This implies a negative value for the coefficient β_2 in the regression (Hetzel's notation; see p. 33, eq.1)

$$(TG_1\text{-}TG_0) = \beta_1 + \beta_2(AG_0\text{-}TG_0), \tag{6.8}$$

where TG_1 is the current target for the growth rate of the money supply, TG_0 is the target of the preceding period, and AG_0 is the actual growth rate of the preceding period.

Table (6.7) presents both Hetzel's and our own regression results. The estimation method is OLS, and the data are quarterly, from 1970.II to 1975.1 (see table A.4 in the appendix).

The table shows that the estimate for β_2 is positive, contrary to conventional wisdom. Table (6.7) also gives regression results when the second and third quarter 1971 are dropped from the data set. We return to this point later. First, we present the results of our tests in table (6.8).

The tests show that Hetzel's refutation of the conventional hypothesis should be greeted with some suspicion. The many rejections of his model are even more surprising in view of the moderate data set (T = 19 observations), where the power for many tests is not very high.

Table 6.7: MONEY GROWTH REGRESSIONS

| | expl. variables | | |
	const.	$AG_0\text{-}TG_0$	d
Hetzel	.01	.32	2.9
	(.02)	(1.8)	
OLS	.01	.32	1.9
	(.02)	(1.8)	
without outl.	.06	.18	1.5
	(.27)	(1.9)	

Table 6.8: TEST RESULTS FOR MONETARY GROWTH REGRESSION

Test	value of test stat.	distr. under H_o	prob-value	rej. at 5 %
Chow (1974.I)	.379	F (2,15)	69.11 %	-
Breusch-Pagan	1.214	χ^2 (2)	27.06 %	x
Breusch-Pagan (s)	.600	χ^2 (2)	43.86 %	x
CUSUM				-
CUSUM (s)				-
CUSUM of sq. (1971.III)				x
CUSUM of sq. (s)	1.036			-
Fluct.	.454	1.480		-
RESET (a)	7.934	F (2,15)	.45 %	x
RESET (b)	1.526	F (1,16)	23.45 %	-
RESET (c)	7.934	F (2,15)	.45 %	x
Harvey-Collier (1)	-3.777	t (16)	.17 %	x
Godfrey-Wickens Lin				n.c.
Rainbow	7.173	F (10,7)	.79 %	x
Outlier (1971.II)	3.808			x
Hausman	8.169	χ^2 (1)	.43 %	x
Diff.	2.824	χ^2 (1)	9.29 %	-
White	7.138	χ^2 (3)	6.76 %	-
IMT	9.708	χ^2 (6)	13.75 %	-

We use this example also to warn against an unreflected routine application of a whole battery of tests, when the same result can be had much cheaper by a simple inspection of the data. Figure (6.1), where the variables of the regression (6.8) are plotted against each other, shows that there is not much of a linear relationship, and that the rejection of the equation by tests for outliers or functional form should not come as a surprise.

The figure clearly identifies the outlier that our formal test also discovered in the second quarter of 1971, and also a second one which happens to be in the third quarter of the same year. Omitting these observations produces the regression results in the bottom of table (6.7). Since the outliers are counteracting each other, the coefficient estimates do not change much after dropping them.

FIG. 6.1: SCATTERPLOT FOR MONETARY GROWTH REGRESSION

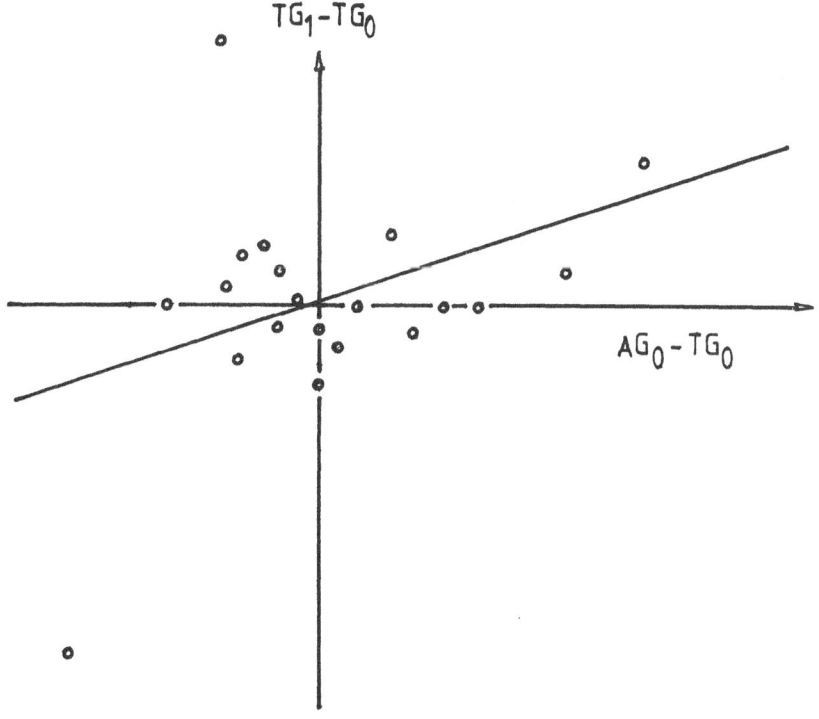

The value of stocks

When estimating a consumption function, many people include some measure of wealth among the explanatory variables. Next we discuss an effort by *Woglom* (1981) to quantify the role of stocks as such a measure of wealth.

We confine ourselves to the first stage of Woglom's analysis, which is to distinguish between the value of stocks as a measure or proxy of consumer sentiment, and the value of stocks as a function of monetary policy. Woglom argues that there are two roles for stocks in the consumption function: (i) the role of stocks as a component of household net worth, and (ii) the role of stock values as an implicit measure of consumer expectations of future earnings, and that different numerical values can be assigned to the value of stocks in each role.

To estimate the value of stocks as a function of monetary policy, Woglom runs the following regression (p.217, eq 2, Woglom's notation):

$$\log VST = \beta_1 + \beta_2 \log MB + \beta_3 \log(1 + RTPD) + \beta_4 \log(1 + RTPS) + \beta_5 \log XBC \ , \ (6.9)$$

where

VST:	value of stocks,
MB:	monetary base,
RTPD:	dollar rent on producer durables,
RTPS:	dollar rent on producer structures,
XBC:	production capacity for business output.

The antilog of the fitted dependent variable from this regression is then taken to be the value of stocks as a function of monetary policy.

Table (6.9) summarizes the estimation results. Woglom reports only results based on a Cochrane-Orcutt transformation, with ϱ as given in the table. The data are quarterly, covering the period 1960.I to 1977.III, and are reproduced in table A.5 in the appendix. Though made available to us by the author via the editorial staff of the Journal of Money, Credit and Banking, they are not identical to those used for Woglom's original regression (the original data set was destroyed and could only partially be reconstructed). This most probably accounts for the discrepancies between our regression results and Woglom's.

Table 6.9: VALUE OF STOCKS REGRESSION

	explanatory variables					d	ϱ
	const.	MB	RTPD	RTPS	XBC		
Woglom	-.9 (-.2)	2.7 (3.8)	1.9 (1.4)	-7.5 (6.2)	-.4 (-.4)	1.3	.78
CO	-2.3 (.5)	2.6 (3.6)	1.9 (1.4)	-7.6 (6.4)	-.1 (.1)	1.4	.78
OLS	1.1 (.6)	2.7 (6.9)	.6 (.7)	-6.2 (7.5)	-.7 (1.4)	.5	

Table 6.10: TEST RESULTS FOR VALUE OF STOCKS REGRESSION

Test	value of test stat.	distr. under H_o	prob-value	rej. at 5 %
		a) CO		
Chow (1974.IV)	3.514	F (5,61)	.74 %	x
Breusch-Pagan	7.545	χ^2 (4)	10.97 %	-
Breusch-Pagan (s)	9.394	χ^2 (4)	5.20 %	-
CUSUM				-
CUSUM (s)				-
CUSUM of sq.				-
CUSUM of sq. (s)	.909			-
Fluct.	13.636	1.620		x
RESET (a)	.346	F (2,64)	70.90 %	-
RESET (b)	.235	F (4,62)	91.73 %	-
RESET (c)	.065	F (2,64)	93.67 %	-
Harvey-Collier (1)	-.506	t (65)	61.48 %	-
Harvey-Collier (2)	-.730	t (65)	46.79 %	-
Harvey-Collier (3)				n.c.
Harvey-Collier (4)	-.743	t (65)	46.02 %	-
Godfrey-Wickens Log	1.358	χ^2 (1)	24.38 %	-
Rainbow	1.038	F (36,30)	46.17 %	-
Outlier (1975.II)	2.470			-
Hausman	16.803	χ^2 (4)	.21 %	x
Diff.				n.c
White	17.153	χ^2 (15)	30.98 %	-
IMT				n.c.
		b) OLS		
Chow (1970.I)	4.481	F (5,61)	.15 %	x
Breusch-Pagan	2.508	χ^2 (4)	64.33 %	-
Breusch-Pagan (s)	3.780	χ^2 (4)	43.65 %	-
CUSUM				-
CUSUM (s)				-
CUSUM of sq.				-
CUSUM of sq. (s)	.974			-
Fluct.	57.929	1.620		x
RESET (a)	1.620	F (2,64)	20.60 %	-
RESET (b)	1.446	F (4,62)	22.97 %	-
RESET (c)	.262	F (2,64)	77.02 %	-
Harvey-Collier (1)	1.746	t (65)	8.55 %	-
Harvey-Collier (2)	1.231	t (65)	22.26 %	-
Harvey-Collier (3)	1.377	t (65)	17.33 %	-
Harvey-Collier (4)	1.718	t (65)	9.06 %	-
Godfrey-Wickens Log	5.000	χ^2 (1)	2.53	x
Rainbow	1.103	F (36,30)	39.48 %	-
Outlier (1966.IV)	2.818			-
Hausman				n.c.
Diff.	2.370	χ^2 (4)	66.81 %	-
White	21.027	χ^2 (15)	13.60 %	-
IMT				n.c.

Table (6.10) summarizes the test results. The presentation follows the familiar pattern, with "n.c." signifying that the respective statistic could not be computed by our software package.

The table shows that equation (6.9) stands up to our challenge, both when estimated before and when estimated after a Cochrane-Orcutt transformation of the data. Both versions fail the Chow and the Fluctuation tests. This is not surprising in case of the Chow test, since the sample was split where a rejection is very likely, i.e where the Quandt-Ratio is smallest. In addition, the model fails the Hausman test when there is a CO-transformation, and the Godfrey-Wickens test for correctness of the log-linear functional form otherwise. The Hausman test could not be computed for the untransformed data.

Wages

The mechanics of wage inflation are among the most debated issues in macroeconomics. Here we examine a contribution by *Nichols* (1983), who models wages in a given labour force group as a function of past inflation, unemployment and the wages of other groups. Among the equations he presents, the first one to meet our criteria concerns low income white collar workers and reads (p.206, table 3, eq. 1)

$$W = 4.28 + .519 \ CPI_{-1} + .121 \ CPI_{-2} + .214 \ CPI_{-3} - .488 \ U + .032 \ MW$$
$$(8.78) \quad (6.04) \qquad\quad (1.04) \qquad\qquad (3.64) \qquad\quad (4.42) \qquad (1.86)$$

$$(6.10)$$

where

W:	wages,
CPI:	consumer price index,
U:	unemployment,
MW:	minimum wage.

Equation (6.10) was estimated by OLS, using annual data from 1962 to 1979. The data were given to us by Donald Nichols and are reproduced in table A.6 in the appendix. Our own estimates agree with (6.10) up to three decimal figures and are for this reason not explicitly reproduced here. This was the only case where so close an agreement between the reported results and our own results could be obtained.

Table (6.11) summarizes the tests. There are several rejections despite the small sample size ($T = 18$) and the very few degrees of freedom ($T-K = 13$), which certainly call for some amendments of the simple equation (6.10).

Table 6.11: TEST RESULTS FOR WAGE REGRESSION

Test	value of test stat.	distr. under H_o	prob-value	rej. at 5 %
Chow (1972)	1.537	F (6,6)	30.75 %	-
Breusch-Pagan	2.053	χ^2 (5)	84.16 %	-
Breusch-Pagan (s)	3.004	χ^2 (5)	69.93 %	-
CUSUM				-
CUSUM (s) (1977)				x
CUSUM of sq.				-
CUSUM of sq. (s)	1.229			-
Fluct.	3.851	1.650		x
RESET (a)	..864	F (2,10)	45.08 %	-
RESET (b)	8.320	F (5,7)	.74 %	x
RESET (c)	2.035	F (2,10)	18.13 %	-
Harvey-Collier (1)	-2.018	t (11)	6.86 %	-
Harvey-Collier (2)	-4.144	t (11)	.16 %	x
Harvey-Collier (3)	-2.204	t (11)	4.97 %	x
Harvey-Collier (4)	-.209	t (11)	83.86 %	-
Harvey-Collier (5)				n.c.
Godfrey-Wickens Lin	5.657	χ^2 (1)	1.74 %	x
Rainbow	.610	F (9,3)	75.16 %	-
Outlier (1972)	2.771			-
Hausman	1.615	χ^2 (3)	89.94 %	-
Diff.	3.139	χ^2 (5)	67.85 %	-
White				n.c.
IMT				n.c.

Unemployment

The Phillips curve is a household word among economists, although many doubt that it really exists. Here we consider a study by *Rea* (1983) who tried to resolve this issue empirically.

His model is (Rea's notation; see p. 185, eq. 7)

$$UN = h_0 + h_1 \log(M/P) + h_2 \log(G) + h_3 \log(X) + h_4 t \qquad (6.11)$$

where

UN:	unemployment rate,
M:	broad money supply,
P:	implicit deflator for Gross National Product,
G:	real purchases of goods and services by Federal, state and local governments,
X:	real exports,
t:	time trend.

Equation (6.11) is one of three models Rea investigates to determine which best explains inflation and unemployment over the 1895-1979 period. It is obtained by expressing an aggregate demand function in terms of the unemployment rate rather than real output. We selected this particular relationship for testing because it is the only one without lagged dependent variables.

Rea views (6.11) as part of a simultaneous equation system and estimates it with full-information maximum likelihood. We disregard any simultaneous equation complications and use OLS, since the only independent variable affected thereby is P, and the coefficient of M/P turns out rather small.

Table 6.12: UNEMPLOYMENT REGRESSION

	explanatory variables				
	const	log(M/P)	log(G)	log(X)	t
Rea	89.79	-0.15	-0.05	-0.09	0.97
	(18.52)	(0.06)	(0.02)	(0.02)	(0.20)
OLS	86.70	-0.13	-0.06	-0.10	0.92
	(9.73)	(0.03)	(0.01)	(0.01)	(0.11)

The estimates are given in table (6.12). Following Rea, they are based on annual data from the period 1895 to 1956. The remaining data are used by Rea to predict unemployment and thus to evaluate the competing models. Also numbers in parentheses now are standard deviations rather than t values. Rea's estimates are those in his trade-off model on p.187, table 1. The data are reproduced in table (A.7) in the appendix.

The estimates for eq. (6.11) are consistent with macroeconomic theory. However, Rea admits that this relationship "was selected after some empirical experimentation" (p. 185, footnote 6). Some call this data mining (of course more authors have probably done this, but no one has openly said so). We are therefore particularly interested in how well the model withstands our tests. The results are in table (6.13).

Table 6.13: TEST RESULTS FOR UNEMPLOYMENT

Test	value of test stat.	distr. under H_o	prob-value	rej. at 5 %
Chow (1941)	2.794	$F_{(5,52)}$	2.62 %	x
Breusch-Pagan	5.717	$\chi^2 (4)$	22.13 %	-
Breusch-Pagan (s)	6.132	$\chi^2 (4)$	18.95 %	
CUSUM				-
CUSUM (s)				-
CUSUM of sq.				-
CUSUM of sq. (s)	1.249			-
Fluct.	15.3o3	1.620		x
RESET (a)	13.570	$F_{(2,55)}$	.oo %	x
RESET (b)	5.300	$F_{(4,53)}$	.11 %	x
RESET (c)	.462	$F_{(2,55)}$	63.26 %	-
Harvey-Collier (1)	.196	$t_{(56)}$	84.5o %	-
Harvey-Collier (2)	-1.746	$t_{(56)}$	8.64 %	-
Harvey-Collier (3)	.743	$t_{(56)}$	46.07 %	-
Harvey-Collier (4)	-1.663	$t_{(56)}$	10.2o %	-
Godfrey-Wickens Log(1,2,3)	.398	$\chi^2 (1)$	52.83 %	-
Rainbow	1.962	$F_{(31,26)}$	4.16 %	x
Outlier (1932)	3.494			x
Hausman	5.862	$\chi^2 (4)$	20.97 %	-
Diff.	1.407	$\chi^2 (4)$	84.30 %	-
White	32.013	$\chi^2 (15)$	.64 %	x
IMT	30.519	$\chi^2 (21)$	8.20 %	-

The tests show that data mining, i.e. searching for significant t-values and correct signs, is no insurance against failure when the tests are more sophisticated. Given the long estimation period, we were surprised that the equation did not fail even more of the tests for structural change. Again, the failure under the Chow test is no surprise, since we let the sample determine where to split the data.

Discussion

Various features of our test results need further clarification. Most important, it is not at all self-evident that the tests retain their (asymptotic) null distribution when applied to the data after a Cochrane-Orcutt transformation. This issue certainly merits further investigation. Meanwhile, we have applied all tests also to the untransformed data, even when the original authors used the transformed ones in their empirical equations, with the result that we usually obtain a lot more failures that way. This could possibly result from the sensitivity of many tests to autocorrelation among the disturbances. Whether or not this is a good thing is not clear. Many might welcome a test that detects more model deviations than it was designed for. On the other hand, failures of the stochastic specification (autocorrelation, heteroskedasticity, nonnormality) have rather negligible effects. Thus one would prefer tests that only point out serious deviations from the model assumptions, such as structural change, incorrect functional form, or omitted variables.

This leads directly to the issue of which conclusion to draw from the results of a given battery of tests, and to testing strategies. We will briefly return to this point below. The present section is mainly concerned with pointing out weaknesses in previous empirical research, and does not propose any remedies.

A related issue concerns the type one error when several tests are applied to identical data sets. Obviously, even the best of models is likely to be rejected at least once when subjected to 20 tests (which is about our average). Given that all tests attain their nominal significance level of 5%, the expected number of failures for any given model is exactly one. Again, we will return to this point in section b.

The test results also allow some inference about the power of the various procedures. For instance, the Fluctuation test rejected more often than other tests for structural change, confirming our Monte Carlo results from chapter 4, and among Hausman-type procedures, the standard Hausman test, where lagged independent variables were used as instruments, outperformed the differencing test (assuming the models are indeed at fault).

b) Issues in Multiple Testing

In section a) we have subjected various given data sets to repeated tests. This raises several questions about the validity of statistical inferences drawn from such procedures. The theoretical literature is rather silent here, despite the practical importance of such issues. Below we review some partial answers which are so far available.

Controlling the size

Assume that n tests are applied to a given model at a time, where the null hypothesis in each case is that assumptions A1-A4 from the Standard Linear Regression Model hold. Disregard any problems related to the asymptotic nature of certain tests and assume that all individual tests exactly attain their nominal significance level α.

Consider first the following rule (which is often called an "induced test"): "Reject the model when at least one individual test is significant". What is the probability of a Type I error for this induced test?

The answer obviously depends on the joint distribution. If nothing is known about this, the best one can hope for is some upper bound for the size of the induced test, which can for instance be found via the Bonferroni inequalities.

Let $C_1, C_2, \dots, C_n$ be the critical regions (rejection regions) of the n individual tests. The critical region of the induced test then is $C = \bigcup_{i=n}^{n} C_i$, and from the principle of inclusion and exclusion we have

$$P(C) = P_1 - P_2 + P_3 - P_4 \pm \dots \pm (-1)^{n-1} P_n , \tag{6.12}$$

where

$$P_1 = \Sigma_{i=1}^{n} P(C_i) \tag{6.13}$$

$$P_2 = \Sigma_{\substack{i,j=1 \\ i<i}}^{n} P(C_i \cap C_j)$$

$$P_3 = \Sigma_{\substack{i,j,l=1 \\ i<j<l}}^{n} P(C_i \cap C_j \cap C_l)$$

$$\vdots$$

$$P_n = P(C_1 \cap C_2 \cap \dots \cap C_n) . \tag{6.14}$$

The sum of the first r terms on the right side of (6.12) provides an upper bound to $P(C)$ when r is odd and a lower bound when r is even. This immediately yields the following sequence of Bonferroni inequalities:

$$P(C) \leq P_1 \tag{6.15}$$

$$P(C) \leq P_1 - P_2 + P_3 \tag{6.16}$$

$$P(C) \leq P_1 - P_2 + P_3 - P_4 + P_5 \tag{6.17}$$

and so on. In practice, only the first inequality (6.15) is of any importance, since the remaining inequalities require the probabilities of joint events, which are usually unknown.

From (6.15), the size of the induced test obeys the trivial inequality

$$P(C) \leq n\alpha .$$ (6.18)

In section a, where we often had up to $n = 20$ individual tests, this implies the rather uninformative result that the probability of at least one rejection at the individual level $\alpha = 5\%$ is less than or equal to one.

An induced test of predetermined size α can of course always be found by ascribing the size α/n to the individual tests, as we did when testing for outliers in section 4.b. The true size of the induced test will then be less than the nominal size (the induced test is conservative), but the margin can be so substantial that the power is also very low.

The size of the induced test can be computed exactly when the individual tests are independent. Then,

$$P(C) = 1 - \prod_{i=1}^{n} P(\bar{C}_i) = 1 - (1-\alpha)^n.$$ (6.19)

In most applications, (6.19) will be closer to the true size of the induced test than P_1 from (6.13). The inequality (6.15) is an equality if and only if the rejection regions of the individual tests are pairwise non-overlapping (up to subsets of probability zero). This means that the individual tests must be extremely negatively correlated in the sense that no two tests can reject at the same time. One can of course always construct such tests, but this requirement will almost never hold for the procedures discussed so far. On the contrary, these tests will in general be positively correlated in the sense that the conditional rejection probability of test A, given test B rejects, exeeds the unconditional probability.

In general, the inequality (6.15) is sharpest when the events C_i and C_j have high negative dependence, and is most off the mark when the dependence is highly positive (i.e. when most of the probability mass of $C = \bigcup C_i$ is concentrated in $\cap C_i$). The most extreme positive dependence occurs when all the rejection regions C_i are identical, in which case $P(C) = \alpha$. As a compromise, we suggest (6.19) rather than (6.15) as a convenient approximation to the true size of the induced test.

Rüger's test

For $\alpha = 5\%$ and $n = 20$, (6.19) still gives a size for the induced test as large as 64%. This means that even a correct model is likely to fail at least once. The obvious solution is to reject a model only if it fails at least k tests, where $k > 1$. This rule has been investigated by *Rüger* (1978).

Let J be the set of all k-element subsets of $\{1,...,n\}$, i.e.

$$J \equiv \{I : I \subset \{1,...,n\}, |I| = k\}. \tag{6.20}$$

The rejection region of Rüger's test is then

$$\tilde{C} = \bigcup_{I \in J} \bigcap_{i \in I} C_i . \tag{6.21}$$

Given k and α, Rüger shows that

$$P(\tilde{C}) \leq \frac{n}{k} \alpha \tag{6.22}$$

under H_0. For $k = 1$, this reduces to the Bonferroni inequality (6.18).

The following simple proof of (6.22) is due to *Morgenstern* (1980):

Let N be the sum of the indicator functions of the individual critical regions C_i $(i = 1,...,n)$, i.e.

$$N(\omega) = \Sigma_{i=1}^{n} I_i(\omega) ,$$

where $I_i(\omega) = 1$ if $\omega \epsilon C_i$ and $I_i(\omega) = 0$ otherwise. Obviously, N is a nonnegative random variable with expectation at most n (under H_0). Therefore, by Markoff's inequality,

$$P(N \geq k) \leq \frac{1}{k} E(X) \leq \frac{n}{k} \alpha , \tag{6.23}$$

proving (6.22).

The inequality (6.22) can be generalized to different significance levels α_i of the individual tests, as follows: Assume that $\alpha_1 \leq \alpha_2 \leq ... \leq \alpha_n$. Along the same lines as above, one proves that

$$P(\tilde{C}) \leq \min_{r=0,...,k-1} \frac{1}{k-r} \Sigma_{i=1}^{n-r} \alpha_i , \tag{6.24}$$

which reduces to (6.22) if $\alpha_1 = \alpha_2 = ... = \alpha_n$.

Assume that $\frac{n}{k} \alpha < 1$. Then one can always find tests such that the equality sign holds in (6.22). This happens whenever the tests are such that either none or exactly k of them are significant. Since this will not often occur in practice, the upper bound (6.22) will in most applications be overly conservative.

Again it is instructive to compare this conservative upper bound to the exact rejection probability under independence. Generalizing (6.19), the probability of at least k failures is

$$P(\tilde{C}) = 1 - \Sigma_{i=0}^{k-1} \binom{n}{i} (1-\alpha)^{n-j} \alpha^j , \tag{6.25}$$

Table (6.14) compares the upper bound (6.22) to the probability (6.25) for various values of k, n and α (Rüger's bounds in parentheses).

Table 6.14: UPPER BOUNDS AND EXACT REJECTION PROBABILITIES

n	k				
	1	2	3	4	5

a) α = 5%

n	1	2	3	4	5
5	22.62(25.00)	2.26(12.50)	0.12(8.33)	0.00(6.25)	0.00(5.00)
10	40.13(50.00)	8.61(25.00)	1.15(16.66)	0.10(12.50)	0.00(10.00)
20	64.15(100.0)	26.42(50.00)	7.55(33.33)	1.59(25.00)	0.26(20.00)

b) α = 1%

n	1	2	3	4	5
5	4.90(5.00)	0.10(2.50)	0.00(1.67)	0.00(1.25)	0.00(1.00)
10	9.56(10.00)	0.43(5.00)	0.01(3.33)	0.00(2.50)	0.00(2.00)
20	18.21(20.00)	1.69(10.00)	0.10(6.67)	0.00(5.00)	0.00(4.00)

Global vs. multiple significance level

So far, the explicit null hypothesis of all tests has been that the assumptions of the Standard Linear Regression Model apply. This includes the requirements of parameter constancy, linearity, and independence of regressors and disturbances, which are the main concern of this monograph.

Many tests however retain their (asymptotic) null distribution even when some model assumptions fail. For instance, the normality assumption for the regression disturbances is usually not needed to establish the asymptotic null distribution of test statistics. A test is then called robust to this particular model deviation, and its implicit null hypothesis includes all models and parameters whose (asymptotic) rejection probabilities equal the size of the test.

Let us therefore briefly consider n tests, with critical regions $C_1, ..., C_n$, where the respective null hypotheses $H_1, ..., H_n$ are different. Then $H = H_1 \cap H_2 \cap ... \cap H_n$ is called the *global* null hypothesis. Assume that H is non-empty. In the present context, H is simply the set of assumptions that constitute the Standard Linear Regression Model.

Consider the induced test where H is rejected whenever at least one H_i (i = 1,...,n) is rejected. The rejection region of this test is $\bigcup_{i=1}^{n} C_i$.

We call

$$\sup_H P(\bigcup C_i)$$

the *global* (or overall) significance level for the induced test. It is this quantity that we have so far sought to control.

Sometimes this may not be enough. Whenever interest focuses on the individual null hypotheses H_i, and on the differences among them, it would be stupid to throw away the information in the individual test results. By the same token, one wants to control the probability of incorrectly rejecting *any* true individual hypothesis H_i. Formally, what is called for is the knowledge of

$$\max_{I \in J} \sup_{\bigcap_{i \in I} P_{H_i}}(\bigcup_{i \in I} C_i) , \qquad (6.26)$$

where J is the class of all non-empty index sets $I \subseteq \{1,2,...,n\}$. The quantity (6.26) is called the *multiple level of significance* for the n individual tests. The vector of n individual tests is called the *joint test* or *multiple test*.

If one views the i'th individual test as the indicator function I_i of the i'th critical region C_i, the multiple test is an n-vector $(I_1, I_2, ..., I_n)$ of binary random variables, whereas the induced test is simply the indicator function of $C = \bigcap C_i$.

Consider as an example the Linear Regression Model, where the possible deviations are restricted to a change, at some known switchpoint T^*, of the regression coefficients β and the disturbance variance σ^2. Let the individual null hypotheses be

$$H_1 : \Delta \beta = 0$$

and

$$H_2 : \sigma_1^2 = \sigma_2^2 .$$

H_1 does not require that the variance is the same for both regimes, and H_2 does not assume constancy of the regression coefficients. The global null hypothesis $H = H_1 \cap H_2$ requires that all assumptions of the Standard Linear Regression Model hold.

Assume that individual tests for each H_i are available, with a global significance level α for the induced test. We have argued in this monograph that a violation of H_2 will in general still allow meaningful inference, whereas a failure of H_1, if unrecognized, will ruin any empirical work. We are therefore also interested in the multiple significance level for the joint test, which in particular gives a bound to the probability of incorrectly rejecting H_1.

The explicit null hypothesis for instance for the Chow test is $H = H_1 \cap H_2$ rather than H_1. Whether or not its *implicit* null hypothesis includes H_1 has been the subject of much debate (see e.g. *Schmidt* and *Sickles*, 1977). We will return to this point later. For the moment, keep in mind that the individual test for H_1 must not be the Chow test, if the multiple significance level is to be controlled.

Obviously, if a joint test has multiple level of significance α, the induced test of $H = \bigcap H_i$ has global significance at most equal to α. The converse need not be true. However, if the (nominal) significance level of the induced test is found via the Bonferroni inequality (6.15), that is by ascribing the significance level α/n to the individual tests, then the joint test will then have a multiple significance level of at most α. This multiple test is also called the Bonferroni test.

Holm's procedure

Assume that the critical regions C_i for the individual tests are based on test statistics S_i $(i = 1,...,n)$. For concreteness, assume in addition that H_i is rejected for large values of S_i. Let $\hat{\alpha}_i$ be the prob-value for the observed value of S_i, that is

$$\hat{\alpha}_i = \sup_{H_i} P(S_i \geq s_i) , \qquad (6.27)$$

where s_i is the sample value of S_i. The multiple Bonferroni test then rejects H_i if $\hat{\alpha}_i \leq \alpha/n$.

Holm (1979) has shown that this procedure can be improved in a very simple way. To see this, let $\hat{\alpha}^{(1)} \leq \hat{\alpha}^{(2)} \leq ... \leq \hat{\alpha}^{(n)}$ be the ordered prob-values, and let $H^{(1)},...,H^{(n)}$ be the corresponding null hypotheses. Holm then proceeds as in figure (6.2) (see also Holm, 1979, scheme 1).

The test proceeds from the top of the scheme until no further rejections can be done. This can happen either by accepting all remaining hypotheses or by rejecting the last hypothesis $H^{(n)}$. This simple sequentially rejective Bonferroni test has multiple level of significance α, which can be seen as follows (Holm, 1979, p.67):

Let J be the index set of the true hypotheses. Let $m = |J|$. Then

$$P(\hat{\alpha}_j > \alpha/m \quad \text{for all } j \epsilon J)$$

$$= 1 - P(\hat{\alpha}_j \leq \alpha/m \quad \text{for some } j \epsilon J)$$

$$\geq 1 - \Sigma_{j \epsilon J} \, P(\hat{\alpha}_j \leq \alpha/m)$$

$$\geq 1 - m\alpha/m = 1 - \alpha . \qquad (6.28)$$

However, the event

$$\{\hat{\alpha}_j > \alpha/m \quad \text{for all } j \epsilon J\}$$

implies the event

$$\{\hat{\alpha}^{(n+1-m)} > \alpha/m \},$$

so the sequentially rejective test stops in step $n+1-m$ or earlier. Consequently, all hypotheses corresponding to prob-values $\hat{\alpha}_i > \alpha/m$ will be accepted, and this set includes the true hypotheses.

FIG. 6.2: THE SEQUENTIALLY REJECTIVE BONFERRONI TEST

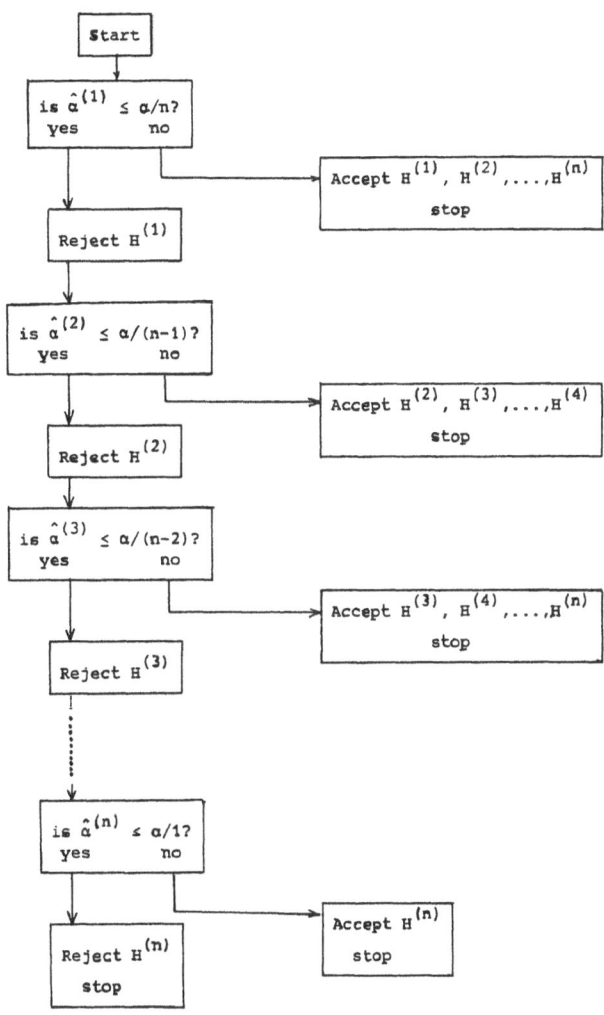

Holm's procedure compares the prob-values of the individual tests to the numbers

$$\alpha/n,\ \alpha/(n\text{-}1),...,\ \alpha/1\ ,$$

whereas the classical Bonferroni test compares all prob-values to α/n. Holm's procedure therefore rejects more often than the Bonferroni test, regardless of which hypotheses are true. In particular, it has higher power.

Obviously, this approach makes sense only when the individual hypotheses H_i are different. Holm's procedure then devides the hypotheses under test into accepted and rejected ones. If all the H_i's are identical and equal to H, and if in addition H is rejected if at least one individual test rejects, then Holm's procedure rejects H if and only if $\hat{\alpha}^{(1)} \leq \alpha/n$, which amounts to the standard Bonferroni test.

Robustness

Multiple testing aggrevates the problem of which conclusion should be drawn when the model is rejected. This problem is present also with a single test, but is then easily wiped under the carpet by narrowing the range of alternatives. With multiple tests, there is no way around the acknowledgement that a rejection can be due to many reasons. It would thus be useful to know which alternatives a given test is robust against, in the sense that it retains its (asymptotic) null distribution also under specified non-ideal conditions.

As an example, consider the Goldfeld-Quandt test for heteroskedasticity of the regression disturbances. The test is obtained by splitting the model $y = X\beta + u$ into $y_1 = X_1\beta_1 + u_1$ and $y_2 = X_2\beta_2 + u_2$, and applying OLS to each subsample separately. The test statistic is

$$ R = \frac{\hat{u}_1{}'\hat{u}_1/(T^*-K)}{\hat{u}_2'\hat{u}_2/(T-T^*-K)} \, , \tag{6.29} $$

where T^* is the size of subsample no. 1, and $\hat{u}_i$ is the OLS residual vector from the i'th regression (i = 1,2). Given assumptions A1-A4 from page 3, R has an F distribution with T^*-K and $T-T^*-K$ degrees of freedom. This distribution is obviously unchanged when $\beta_1 \neq \beta_2$. The Goldfeld-Quandt test is thus perfectly robust to changes in the regression coefficients. This implies that a significant Goldfeld-Quandt test unambiguously points towards a change in variance (given that the only possible violations of the model assumptions are $\beta_1 \neq \beta_2$ and $\sigma_1^2 \neq \sigma_2^2$).

By contrast, the Chow test, which is designed to detect changes in β, is not robust to changes in σ^2. In particular, rejections can occur with high probability due to $\sigma_1^2 \neq \sigma_2^2$, even if $\beta_1 = \beta_2$. A significant Chow test thus does not give an unambiguous answer, even if the range of possible alternatives is narrowed down to $\beta_1 \neq \beta_2$ and $\sigma_1^2 \neq \sigma_2^2$.

So far, this issue of robustness has been addressed only for selected special cases. *Jayatissa* (1977) has developed an alternative to the Chow test that is robust to heteroskedasticity. *Epps* and *Epps* (1977) have shown that the Durbin-Watson test is fairly robust to heteroskedasticity, as opposed to a high sensitivity to autocorrelation of the Goldfeld-Quandt test. *White* (1980) has shown how to modify t- and F-tests such as to be robust against heteroskedasticity, and *Thursby* (1981, 1982) concludes from Monte Carlo experiments that RESET tests are robust to both heteroskedasticity and autocorrelation (a claim shown to be not generally valid by *Pagan*, 1984). Most of the procedures considered in this monograph are also known to be asymptotically robust to nonnormality. Given however the enormous variety of possible violations of the ideal model assumptions, and the notorious non-robustness of many tests to most of them, one cannot reasonably hope for easy answers, except by assuming troublesome alternatives away. In particular, there is no way to distinguish between different causes of non-zero means of the disturbances, as has been demonstrated by *Thursby* (1982) for the problems of omitted variables and structural change.

Summing up, multiple testing is a research activity with many unsolved problems and few solutions. Still, it is better than not to test at all, or in the words of Denis Sargan (1975):

"Despite the problems associated with data mining, I consider that a suggested specification should be tested in all possible ways and only those which survive and correspond to a reasonable model should be used."

Further reading

The classical textbook on simultaneous statistical inference is *Miller* (1966). *Miller* (1977) provides a survey of the more recent literature. Induced tests were first discussed by *Roy* (1953). A general textbook treatment of Bonferroni inequalities appears in *David* (1981), and an up-to-date discussion with many further references is found in *Schwager* (1984). *Sonnemann* (1982) discusses links between Bonferroni tests and Holm's procedure, and *Savin* (1984) investigates in detail the relationship between multiple t-tests and the F-test.

APPENDIX

a) Sample Data

Table A.1: MONEY DEMAND

	I	log M	I	log y_p	I	r_s	I	r_l	I	r_m	Ilog S(p/P)
1879	I	-7.424	I	5.608	I	5.067	I	4.880	I	2.580	I	-3.759
1880	I	-7.253	I	5.695	I	5.230	I	4.580	I	2.730	I	-2.685
1881	I	-7.087	I	5.744	I	5.200	I	4.260	I	2.870	I	-2.703
1882	I	-7.044	I	5.789	I	5.640	I	4.310	I	3.140	I	-2.671
1883	I	-7.006	I	5.810	I	5.620	I	4.330	I	3.160	I	-2.790
1884	I	-7.014	I	5.826	I	5.200	I	4.280	I	2.890	I	-2.723
1885	I	-6.980	I	5.832	I	4.060	I	4.080	I	2.230	I	-3.242
1886	I	-6.894	I	5.848	I	4.760	I	3.810	I	2.860	I	-3.229
1887	I	-6.839	I	5.860	I	5.750	I	3.870	I	3.510	I	-3.330
1888	I	-6.828	I	5.852	I	4.890	I	3.800	I	2.950	I	-3.154
1889	I	-6.790	I	5.852	I	4.860	I	3.660	I	3.020	I	-3.313
1890	I	-6.720	I	5.869	I	5.620	I	3.780	I	3.620	I	-3.863
1891	I	-6.690	I	5.889	I	5.390	I	3.950	I	3.450	I	-3.897
1892	I	-6.613	I	5.929	I	4.110	I	3.830	I	2.600	I	-3.697
1893	I	-6.665	I	5.927	I	6.790	I	3.930	I	4.110	I	-3.654
1894	I	-6.656	I	5.905	I	3.040	I	3.720	I	1.710	I	-3.376
1895	I	-6.619	I	5.928	I	3.670	I	3.590	I	2.250	I	-3.335
1896	I	-6.633	I	5.927	I	5.810	I	3.630	I	3.370	I	-3.321
1897	I	-6.571	I	5.953	I	3.500	I	3.440	I	2.170	I	-3.378
1898	I	-6.464	I	5.967	I	3.820	I	3.380	I	2.460	I	-3.341
1899	I	-6.343	I	6.006	I	4.150	I	3.240	I	2.860	I	-3.547
1900	I	-6.301	I	6.036	I	4.380	I	3.300	I	3.060	I	-3.300
1901	I	-6.211	I	6.092	I	4.280	I	3.250	I	3.080	I	-3.378
1902	I	-6.163	I	6.123	I	4.920	I	3.300	I	3.640	I	-3.448
1903	I	-6.136	I	6.154	I	5.470	I	3.450	I	4.080	I	-3.480
1904	I	-6.108	I	6.162	I	4.200	I	3.600	I	3.100	I	-3.709
1905	I	-6.046	I	6.185	I	4.400	I	3.500	I	3.320	I	-4.220
1906	I	-6.004	I	6.237	I	5.680	I	3.550	I	4.330	I	-4.711
1907	I	-6.004	I	6.267	I	6.340	I	3.800	I	4.830	I	-4.398
1908	I	-6.054	I	6.241	I	4.370	I	3.950	I	3.200	I	-4.167
1909	I	-5.994	I	6.263	I	3.980	I	3.770	I	2.970	I	-4.098
1910	I	-5.987	I	6.273	I	5.010	I	3.800	I	3.790	I	-4.057
1911	I	-5.960	I	6.281	I	4.030	I	3.900	I	3.070	I	-3.808
1912	I	-5.930	I	6.298	I	4.740	I	3.900	I	3.670	I	-3.794
1913	I	-5.925	I	6.318	I	5.580	I	4.000	I	4.340	I	-3.858
1914	I	-5.920	I	6.292	I	4.790	I	4.100	I	3.720	I	-3.917
1915	I	-5.884	I	6.278	I	3.450	I	4.150	I	2.680	I	-3.878
1916	I	-5.786	I	6.319	I	3.420	I	4.050	I	2.700	I	-3.024
1917	I	-5.760	I	6.331	I	4.740	I	4.050	I	3.850	I	-2.362
1918	I	-5.805	I	6.368	I	5.870	I	4.750	I	4.910	I	-2.366
1919	I	-5.743	I	6.404	I	5.420	I	4.750	I	4.480	I	-2.355
1920	I	-5.743	I	6.406	I	7.370	I	5.100	I	5.950	I	-2.383
1921	I	-5.827	I	6.387	I	6.530	I	5.170	I	5.260	I	-1.924
1922	I	-5.804	I	6.392	I	4.420	I	4.710	I	3.560	I	-2.032
1923	I	-5.739	I	6.438	I	4.970	I	4.610	I	4.010	I	-2.087
1924	I	-5.700	I	6.471	I	3.900	I	4.660	I	3.120	I	-2.079

Table A.1 (continued)

	I	log M	I	log yp	I	rs	I	r1	I	rm	Ilog S(p/P)
	I		I		I		I		I		I
1925	I	-5.632	I	6.433	I	4.000	I	4.500	I	3.240	I	-2.146
1926	I	-5.611	I	6.530	I	4.230	I	4.400	I	3.440	I	-3.004
1927	I	-5.594	I	6.552	I	4.020	I	4.300	I	3.270	I	-3.785
1928	I	-5.567	I	6.567	I	4.840	I	4.050	I	4.020	I	-3.979
1929	I	-5.572	I	6.594	I	5.780	I	4.420	I	4.820	I	-3.963
1930	I	-5.584	I	6.571	I	3.550	I	4.400	I	2.800	I	-3.623
1931	I	-5.609	I	6.525	I	2.630	I	4.100	I	1.780	I	-2.851
1932	I	-5.712	I	6.434	I	2.720	I	4.700	I	1.770	I	-2.620
1933	I	-5.775	I	6.357	I	1.670	I	4.150	I	.586	I	-2.675
1934	I	-5.699	I	6.330	I	.880	I	3.990	I	.430	I	-2.467
1935	I	-5.559	I	6.362	I	.750	I	3.500	I	.563	I	-2.483
1936	I	-5.459	I	6.423	I	.750	I	3.200	I	.551	I	-2.606
1937	I	-5.419	I	6.491	I	.940	I	3.080	I	.671	I	-2.915
1938	I	-5.433	I	6.504	I	.860	I	3.000	I	.596	I	-2.888
1939	I	-5.360	I	6.536	I	.720	I	2.750	I	.467	I	-3.291
1940	I	-5.258	I	6.586	I	.810	I	2.700	I	.501	I	-3.352
1941	I	-5.171	I	6.655	I	.700	I	2.650	I	.448	I	-3.294
1942	I	-5.115	I	6.750	I	.690	I	2.650	I	.457	I	-2.832
1943	I	-4.980	I	6.831	I	.720	I	2.650	I	.556	I	-2.709
1944	I	-4.901	I	6.882	I	.750	I	2.600	I	.582	I	-2.766
1945	I	-4.808	I	6.899	I	.750	I	2.550	I	.588	I	-2.886
1946	I	-4.774	I	6.880	I	.810	I	2.430	I	.626	I	-2.822
1947	I	-4.813	I	6.853	I	1.030	I	2.500	I	.729	I	-2.974
1948	I	-4.884	I	6.846	I	1.440	I	2.800	I	1.016	I	-3.265
1949	I	-4.947	I	6.832	I	1.490	I	2.740	I	1.058	I	-2.968
1950	I	-4.972	I	6.847	I	1.450	I	2.580	I	1.056	I	-2.879
1951	I	-4.997	I	6.878	I	2.160	I	2.670	I	1.564	I	-3.018
1952	I	-4.997	I	6.901	I	2.330	I	3.000	I	1.703	I	-3.242
1953	I	-5.003	I	6.926	I	2.520	I	3.150	I	1.853	I	-3.358
1954	I	-5.008	I	6.930	I	1.580	I	3.000	I	1.178	I	-3.673
1955	I	-5.010	I	6.952	I	2.180	I	3.040	I	1.654	I	-3.726
1956	I	-5.034	I	6.965	I	3.310	I	3.090	I	2.511	I	-3.917
1957	I	-5.056	I	6.971	I	3.810	I	3.680	I	2.905	I	-4.276
1958	I	-5.054	I	6.965	I	2.460	I	3.610	I	1.904	I	-4.528
1959	I	-5.052	I	6.978	I	3.970	I	4.100	I	3.114	I	-4.605
1960	I	-5.085	I	6.988	I	3.850	I	4.550	I	3.053	I	-4.636
1961	I	-5.070	I	6.996	I	2.970	I	4.220	I	2.387	I	-4.605
1962	I	-5.046	I	7.019	I	3.260	I	4.420	I	2.653	I	-5.051
1963	I	-5.018	I	7.044	I	3.550	I	4.160	I	2.907	I	-5.006
1964	I	-4.986	I	7.074	I	3.970	I	4.330	I	3.245	I	-4.976
1965	I	-4.938	I	7.111	I	4.380	I	4.350	I	3.600	I	-4.948
1966	I	-4.896	I	7.153	I	5.550	I	4.750	I	4.554	I	-4.976
1967	I	-4.856	I	7.185	I	5.100	I	4.950	I	4.188	I	-4.768
1968	I	-4.814	I	7.219	I	5.900	I	5.930	I	4.870	I	-4.431
1969	I	-4.800	I	7.246	I	7.830	I	6.540	I	6.485	I	-4.186
1970	I	-4.816	I	7.256	I	7.720	I	7.600	I	6.392	I	-4.017
1971	I	-4.760	I	7.271	I	5.110	I	7.120	I	4.261	I	-4.080
1972	I	-4.711	I	7.301	I	4.690	I	7.010	I	3.933	I	-4.186
1973	I	-4.674	I	7.338	I	8.150	I	7.200	I	.000	I	-4.241
1974	I	-4.667	I	7.349	I	9.870	I	7.800	I	.020	I	-2.817
	I		I		I		I		I		I

Table A.2: CURRENCY SUBSTITUTION

	log C$/US$	i_u	i_c	log y
1960Q4	2.662	.022	.033	10.889
1961Q1	2.869	.024	.032	10.874
1961Q2	2.795	.023	.026	10.907
1961Q3	2.609	.023	.026	10.922
1961Q4	2.724	.026	.030	10.938
1962Q1	2.701	.028	.031	10.967
1962Q2	2.426	.029	.055	10.960
1962Q3	2.648	.028	.050	10.982
1962Q4	2.705	.029	.039	10.996
1963Q1	2.729	.030	.036	11.005
1963Q2	2.711	.030	.032	11.013
1963Q3	2.655	.035	.036	11.027
1963Q4	2.726	.036	.038	11.061
1964Q1	2.824	.036	.039	11.076
1964Q2	2.663	.036	.036	11.082
1964Q3	2.541	.036	.037	11.098
1964Q4	2.381	.040	.038	11.110
1965Q1	2.532	.040	.036	11.133
1965Q2	2.676	.039	.039	11.146
1965Q3	2.600	.041	.041	11.161
1965Q4	2.559	.046	.045	11.183
1966Q1	2.462	.047	.051	11.212
1966Q2	2.433	.046	.050	11.222
1966Q3	2.347	.056	.050	11.223
1966Q4	2.380	.049	.050	11.236
1967Q1	2.520	.042	.041	11.243
1967Q2	2.574	.035	.043	11.258
1967Q3	2.528	.047	.048	11.256
1967Q4	2.335	.051	.059	11.268
1968Q1	2.513	.053	.070	11.280
1968Q2	2.431	.054	.066	11.305
1968Q3	2.419	.053	.057	11.322
1968Q4	2.426	.064	.062	11.343
1969Q1	2.401	.061	.066	11.356
1969Q2	2.112	.067	.071	11.352
1969Q3	1.979	.075	.078	11.367
1969Q4	1.995	.084	.078	11.384
1970Q1	2.024	.065	.070	11.383
1970Q2	2.182	.068	.059	11.386
1970Q3	2.102	.060	.054	11.396
1970Q4	2.148	.050	.044	11.393
1971Q1	2.417	.036	.032	11.422
1971Q2	2.516	.052	.034	11.445
1971Q3	2.698	.048	.041	11.473
1971Q4	2.878	.038	.032	11.482

Table A.2 (continued)

I	Ilog C$/US$ I	iu I	ic I	log y I
I 1972Q1 I	3.054 I	.039 I	.036 I	11.489 I
I 1972Q2 I	3.170 I	.042 I	.035 I	11.514 I
I 1972Q3 I	3.240 I	.048 I	.036 I	11.520 I
I 1972Q4 I	3.001 I	.053 I	.036 I	11.539 I
I 1973Q1 I	3.051 I	.064 I	.045 I	11.574 I
I 1973Q2 I	2.888 I	.075 I	.055 I	11.578 I
I 1973Q3 I	2.677 I	.076 I	.065 I	11.587 I
I 1973Q4 I	2.670 I	.077 I	.064 I	11.613 I
I 1974Q1 I	2.487 I	.086 I	.065 I	11.627 I
I 1974Q2 I	2.132 I	.081 I	.088 I	11.622 I
I 1974Q3 I	2.151 I	.066 I	.089 I	11.621 I
I 1974Q4 I	2.398 I	.073 I	.071 I	11.622 I
I 1975Q1 I	2.604 I	.057 I	.063 I	11.625 I
I 1975Q2 I	2.562 I	.058 I	.070 I	11.630 I
I 1975Q3 I	2.680 I	.065 I	.084 I	11.640 I
I 1975Q4 I	2.610 I	.053 I	.086 I	11.650 I

Table A.3: BOND YIELD

	I RAARUS I	MOOD I	EPI I	EXP I	RUS I	Y I	K I
1961Q1 I	.290 I	91.100 I	98.840 I	.460 I	4.010 I	-2.667 I	3.940
1961Q2 I	.730 I	92.300 I	98.500 I	.420 I	3.940 I	-3.238 I	3.936
1961Q3 I	.390 I	93.400 I	98.470 I	.490 I	4.210 I	-.534 I	3.913
1961Q4 I	.250 I	94.400 I	98.820 I	.440 I	4.280 I	-3.710 I	3.969
1962Q1 I	.180 I	97.200 I	98.960 I	.400 I	4.280 I	-2.140 I	3.937
1962Q2 I	.110 I	95.400 I	99.010 I	.430 I	4.170 I	-3.189 I	3.960
1962Q3 I	.240 I	91.600 I	99.300 I	.420 I	4.080 I	-2.666 I	3.951
1962Q4 I	.310 I	95.000 I	99.040 I	.410 I	3.940 I	-1.598 I	3.934
1963Q1 I	.270 I	94.800 I	98.910 I	.480 I	3.960 I	-2.123 I	3.944
1963Q2 I	.310 I	91.400 I	98.970 I	.450 I	4.030 I	-2.127 I	3.947
1963Q3 I	.320 I	96.200 I	98.990 I	.230 I	4.040 I	-2.125 I	3.943
1963Q4 I	.290 I	96.900 I	99.140 I	.170 I	4.150 I	-.526 I	3.930
1964Q1 I	.230 I	99.000 I	99.120 I	.120 I	4.200 I	-1.051 I	3.940
1964Q2 I	.300 I	98.100 I	99.550 I	.150 I	4.160 I	.000 I	3.927
1964Q3 I	.230 I	100.200 I	99.310 I	.190 I	4.210 I	.000 I	3.928
1964Q4 I	.230 I	99.400 I	99.200 I	.070 I	4.210 I	-1.053 I	3.938
1965Q1 I	.210 I	101.500 I	99.360 I	.080 I	4.230 I	.525 I	3.920

Table A.3 (continued)

	I RAARUS I	MOOD	I	EPI	I	EXP	I	RUS	I	Y	I	K
1965Q2 I	.310 I	102.200	I	99.800	I	.110	I	4.230	I	.525	I	3.916
1965Q3 I	.360 I	103.200	I	100.070	I	.090	I	4.290	I	.524	I	3.917
1965Q4 I	.310 I	102.900	I	100.390	I	.010	I	4.460	I	1.556	I	3.915
1966Q1 I	.280 I	100.000	I	100.520	I	.050	I	4.780	I	3.567	I	3.908
1966Q2 I	.490 I	95.700	I	100.680	I	.000	I	4.880	I	2.538	I	3.926
1966Q3 I	.770 I	91.200	I	100.990	I	-.110	I	5.080	I	5.489	I	3.889
1966Q4 I	.940 I	88.300	I	101.310	I	-.150	I	4.820	I	4.037	I	3.907
1967Q1 I	.640 I	92.200	I	101.020	I	.000	I	4.680	I	2.059	I	3.916
1967Q2 I	.690 I	94.900	I	100.550	I	.120	I	5.110	I	-.508	I	3.982
1967Q3 I	.660 I	96.500	I	101.060	I	-.020	I	5.420	I	2.973	I	3.946
1967Q4 I	.550 I	92.900	I	101.230	I	-.010	I	6.010	I	4.806	I	3.949
1968Q1 I	.650 I	95.000	I	101.160	I	.010	I	5.880	I	4.352	I	3.946
1968Q2 I	.990 I	92.400	I	101.390	I	-.080	I	5.820	I	5.268	I	3.934
1968Q3 I	.810 I	92.900	I	100.970	I	-.010	I	5.570	I	2.951	I	3.959
1968Q4 I	.630 I	92.100	I	100.900	I	-.060	I	6.190	I	.968	I	4.005
1969Q1 I	.370 I	95.100	I	101.490	I	-.030	I	6.860	I	.000	I	4.057
1969Q2 I	.790 I	91.600	I	101.210	I	-.080	I	6.840	I	5.111	I	3.966
1969Q3 I	.910 I	88.400	I	101.550	I	-.310	I	7.190	I	6.796	I	3.947
1969Q4 I	1.170 I	79.700	I	101.740	I	-.280	I	7.530	I	8.860	I	3.925
1970Q1 I	1.220 I	78.100	I	101.450	I	-.180	I	7.460	I	14.407	I	3.840
1970Q2 I	1.090 I	75.400	I	100.530	I	-.070	I	8.090	I	11.904	I	3.880
1970Q3 I	.950 I	77.100	I	99.890	I	.060	I	7.750	I	9.481	I	3.920
1970Q4 I	1.000 I	75.400	I	99.280	I	.310	I	7.380	I	8.185	I	3.944
1971Q1 I	.670 I	78.200	I	98.880	I	.580	I	6.740	I	.932	I	4.039
1971Q2 I	1.080 I	81.600	I	98.390	I	.270	I	6.950	I	-.467	I	4.067
1971Q3 I	1.060 I	82.400	I	98.550	I	.250	I	6.790	I	14.581	I	3.824
1971Q4 I	.970 I	82.200	I	98.900	I	.480	I	6.490	I	2.359	I	4.008
1972Q1 I	.710 I	87.500	I	99.380	I	.380	I	6.700	I	6.920	I	3.938
1972Q2 I	.830 I	89.300	I	99.600	I	.310	I	6.650	I	4.167	I	3.985
1972Q3 I	.970 I	94.000	I	99.500	I	.220	I	6.560	I	10.628	I	3.867
1972Q4 I	.930 I	90.800	I	99.530	I	.170	I	6.430	I	15.985	I	3.784
1973Q1 I	.590 I	80.800	I	100.420	I	-.010	I	6.910	I	-8.942	I	4.250
1973Q2 I	.600 I	76.000	I	100.820	I	.000	I	7.080	I	-5.317	I	4.182
1973Q3 I	.670 I	71.800	I	100.940	I	-.180	I	7.410	I	-7.646	I	4.250
1973Q4 I	.610 I	75.700	I	101.500	I	-.020	I	7.320	I-13.204	I	4.340
1974Q1 I	.640 I	60.900	I	101.910	I	-.030	I	7.740	I	-9.959	I	4.286
1974Q2 I	.750 I	72.000	I	101.510	I	-.170	I	8.300	I-14.556	I	4.391
1974Q3 I	1.110 I	64.500	I	101.270	I	-.120	I	8.710	I-12.667	I	4.359
1974Q4 I	1.030 I	58.400	I	100.030	I	.020	I	8.070	I-14.254	I	4.379
1975Q1 I	.950 I	58.000	I	98.300	I	.420	I	8.040	I-20.519	I	4.494
1975Q2 I	.920 I	72.900	I	97.900	I	.210	I	8.350	I-19.009	I	4.482
1975Q3 I	.840 I	75.800	I	98.330	I	.090	I	8.570	I-16.148	I	4.420
1975Q4 I	.680 I	75.800	I	98.330	I	.380	I	8.280	I	-7.305	I	4.251

Table A.4: GROWTH OF MONEY SUPPLY

```
====================================
I          I  TG1-TG0  I  AG0-TG0  I
====================================
I          I          I          I
I 1970Q2   I    .000  I    -.400  I
I 1970Q3   I   1.000  I   -1.000  I
I 1970Q4   I   1.000  I    1.100  I
I 1971Q1   I   2.500  I    5.800  I
I 1971Q2   I  -6.000  I   -4.400  I
I 1971Q3   I   4.500  I   -1.600  I
I 1971Q4   I   -.500  I    1.600  I
I 1972Q1   I  -1.000  I   -1.400  I
I 1972Q2   I    .500  I    4.400  I
I 1972Q3   I  -1.500  I     .000  I
I 1972Q4   I   -.250  I    -.700  I
I 1973Q1   I   -.750  I     .100  I
I 1973Q2   I    .000  I     .650  I
I 1973Q3   I    .750  I   -1.400  I
I 1973Q4   I    .000  I    2.800  I
I 1974Q1   I    .000  I    2.150  I
I 1974Q2   I    .500  I    -.850  I
I 1974Q3   I    .250  I   -1.650  I
I 1974Q4   I    .000  I   -2.550  I
I          I          I          I
====================================
```

Table A.5: VALUE OF STOCKS

```
=================================================================================
I          I     VST    I    MB     I   RTPD    I   RTPS    I    XBC     I
=================================================================================
I          I           I           I           I           I            I
I 1960Q1   I  388.160  I   43.867  I   21.175  I   12.542  I   758.890  I
I 1960Q2   I  383.810  I   43.933  I   20.904  I   12.277  I   764.010  I
I 1960Q3   I  390.000  I   44.033  I   20.517  I   12.059  I   769.280  I
I 1960Q4   I  385.360  I   44.067  I   20.584  I   11.972  I   774.860  I
I 1961Q1   I  429.950  I   44.700  I   20.575  I   11.908  I   780.680  I
I 1961Q2   I  463.100  I   44.667  I   20.394  I   11.944  I   786.610  I
I 1961Q3   I  475.730  I   45.133  I   20.296  I   12.159  I   792.490  I
I 1961Q4   I  507.700  I   45.800  I   20.333  I   12.163  I   798.160  I
I 1962Q1   I  494.540  I   46.100  I   20.447  I   12.144  I   803.680  I
I 1962Q2   I  445.900  I   46.567  I   20.228  I   12.038  I   809.150  I
I 1962Q3   I  422.550  I   46.900  I   18.838  I   12.089  I   813.620  I
I 1962Q4   I  435.540  I   47.400  I   18.540  I   12.025  I   820.210  I
I 1963Q1   I  481.900  I   47.933  I   18.271  I   11.967  I   826.030  I
I 1963Q2   I  513.770  I   48.500  I   18.264  I   12.034  I   832.170  I
I 1963Q3   I  524.210  I   49.167  I   18.270  I   12.121  I   838.600  I
I 1963Q4   I  541.550  I   49.867  I   18.190  I   12.238  I   845.270  I
I 1964Q1   I  573.510  I   50.567  I   17.222  I   12.125  I   852.130  I
I 1964Q2   I  600.880  I   51.167  I   17.156  I   12.200  I   859.190  I
```

Table A.5 (continued)

I 1964Q3 I	616.330 I	52.000 I	16.763 I	12.280 I	866.430 I
I 1964Q4 I	636.980 I	52.733 I	16.413 I	12.344 I	873.870 I
I 1965Q1 I	644.950 I	53.400 I	16.271 I	12.192 I	881.540 I
I 1965Q2 I	672.730 I	54.033 I	16.196 I	12.295 I	889.480 I
I 1965Q3 I	677.610 I	54.767 I	16.038 I	12.493 I	897.650 I
I 1965Q4 I	713.300 I	55.833 I	16.625 I	12.752 I	906.070 I
I 1966Q1 I	689.860 I	56.633 I	16.865 I	13.139 I	914.730 I
I 1966Q2 I	638.380 I	57.467 I	16.861 I	13.369 I	923.670 I
I 1966Q3 I	576.480 I	58.133 I	17.090 I	14.129 I	932.940 I
I 1966Q4 I	546.760 I	58.467 I	17.260 I	14.355 I	942.620 I
I 1967Q1 I	620.830 I	59.333 I	17.206 I	14.170 I	952.680 I
I 1967Q2 I	676.000 I	60.200 I	17.902 I	14.502 I	963.120 I
I 1967Q3 I	711.710 I	61.167 I	18.466 I	15.145 I	973.840 I
I 1967Q4 I	680.900 I	62.233 I	18.767 I	15.874 I	984.700 I
I 1968Q1 I	694.470 I	63.267 I	18.992 I	17.038 I	995.600 I
I 1968Q2 I	758.470 I	64.100 I	19.179 I	17.463 I	1006.460 I
I 1968Q3 I	790.570 I	65.167 I	19.020 I	17.328 I	1017.620 I
I 1968Q4 I	821.210 I	66.467 I	18.762 I	17.820 I	1027.890 I
I 1969Q1 I	764.070 I	67.400 I	19.170 I	18.929 I	1038.540 I
I 1969Q2 I	774.660 I	67.933 I	21.262 I	19.649 I	1049.250 I
I 1969Q3 I	728.400 I	68.500 I	21.295 I	20.488 I	1060.070 I
I 1969Q4 I	731.330 I	69.433 I	22.234 I	21.661 I	1071.030 I
I 1970Q1 I	688.540 I	70.200 I	22.759 I	21.865 I	1082.090 I
I 1970Q2 I	619.630 I	71.333 I	22.977 I	22.685 I	1093.210 I
I 1970Q3 I	616.420 I	72.667 I	23.369 I	23.238 I	1104.290 I
I 1970Q4 I	677.960 I	74.000 I	23.231 I	23.193 I	1115.170 I
I 1971Q1 I	790.240 I	75.633 I	22.348 I	22.036 I	1125.710 I
I 1971Q2 I	827.070 I	77.100 I	22.604 I	22.868 I	1135.940 I
I 1971Q3 I	809.210 I	78.667 I	21.336 I	23.384 I	1145.760 I
I 1971Q4 I	782.700 I	79.600 I	20.415 I	23.434 I	1155.240 I
I 1972Q1 I	902.740 I	81.000 I	20.536 I	23.672 I	1164.470 I
I 1972Q2 I	941.480 I	82.633 I	21.437 I	23.922 I	1173.590 I
I 1972Q3 I	975.890 I	84.100 I	22.625 I	24.031 I	1182.670 I
I 1972Q4 I	1032.430 I	86.233 I	22.959 I	24.233 I	1191.880 I
I 1973Q1 I	1073.480 I	88.233 I	23.594 I	24.919 I	1201.310 I
I 1973Q2 I	1026.740 I	89.967 I	23.672 I	25.644 I	1211.170 I
I 1973Q3 I	1017.640 I	91.733 I	23.720 I	26.916 I	1221.520 I
I 1973Q4 I	980.890 I	93.233 I	22.602 I	27.764 I	1232.510 I
I 1974Q1 I	916.480 I	95.367 I	22.855 I	29.595 I	1244.120 I
I 1974Q2 I	883.770 I	97.567 I	23.125 I	32.575 I	1256.300 I
I 1974Q3 I	728.990 I	99.467 I	24.159 I	36.320 I	1268.940 I
I 1974Q4 I	669.230 I	101.633 I	25.913 I	37.954 I	1281.730 I
I 1975Q1 I	775.100 I	103.233 I	25.865 I	38.451 I	1294.450 I
I 1975Q2 I	885.800 I	105.167 I	27.534 I	39.286 I	1306.870 I
I 1975Q3 I	884.490 I	107.433 I	26.950 I	39.829 I	1318.680 I
I 1975Q4 I	899.700 I	109.400 I	27.114 I	39.871 I	1329.540 I
I 1976Q1 I	1042.540 I	111.533 I	27.401 I	39.331 I	1339.380 I
I 1976Q2 I	1069.650 I	114.333 I	28.380 I	39.767 I	1348.410 I
I 1976Q3 I	1117.070 I	116.300 I	28.662 I	39.701 I	1356.930 I
I 1976Q4 I	1117.720 I	118.633 I	28.529 I	39.402 I	1365.280 I
I 1977Q1 I	1069.120 I	120.800 I	28.730 I	39.942 I	1373.730 I
I 1977Q2 I	1041.160 I	123.333 I	29.074 I	40.673 I	1382.490 I
I 1977Q3 I	1034.330 I	126.233 I	29.675 I	40.780 I	1391.730 I
I	I	I	I	I	I

Table A.6: WAGES

I		I	W	I	CPI	I	U	I	MW	I
I		I		I		I		I		I
I	1960	I	.	I	.347	I	.	I	.	I
I	1961	I	.	I	1.499	I	.	I	.	I
I	1962	I	2.800	I	1.477	I	6.457	I	15.000	I
I	1963	I	2.700	I	1.120	I	6.192	I	.000	I
I	1964	I	2.700	I	1.107	I	5.763	I	8.696	I
I	1965	I	2.200	I	1.424	I	5.018	I	.000	I
I	1966	I	2.900	I	1.188	I	4.170	I	.000	I
I	1967	I	4.500	I	2.775	I	3.725	I	12.000	I
I	1968	I	5.100	I	2.700	I	3.723	I	14.286	I
I	1969	I	5.500	I	3.943	I	3.686	I	.000	I
I	1970	I	6.200	I	5.058	I	3.556	I	.000	I
I	1971	I	6.200	I	6.019	I	5.395	I	.000	I
I	1972	I	6.300	I	4.629	I	6.272	I	.000	I
I	1973	I	5.500	I	3.506	I	5.433	I	.000	I
I	1974	I	6.200	I	4.677	I	4.522	I	.000	I
I	1975	I	9.100	I	10.247	I	6.163	I	31.250	I
I	1976	I	7.600	I	10.273	I	8.625	I	9.524	I
I	1977	I	6.900	I	6.147	I	7.877	I	.000	I
I	1978	I	7.500	I	6.388	I	6.679	I	15.217	I
I	1979	I	7.200	I	6.510	I	5.494	I	9.434	I
I		I		I		I		I		I

Table A.7: UNEMPLOYMENT

```
==============================================================================
      I   UN    I    M      I    P    I   G      I    X     I    T
==============================================================================
      I         I           I         I          I          I
1890  I  3.970  I   3.920  I  .168  I  16.364  I   5.900  I   1.000
1891  I  5.419  I   4.080  I  .166  I  16.619  I   6.300  I   2.000
1892  I  3.045  I   4.430  I  .161  I  16.930  I   6.900  I   3.000
1893  I 11.684  I   4.260  I  .163  I  17.178  I   6.300  I   4.000
1894  I 18.415  I   4.280  I  .155  I  17.364  I   6.400  I   5.000
1895  I 13.703  I   4.430  I  .153  I  17.449  I   6.000  I   6.000
1896  I 14.445  I   4.350  I  .150  I  17.656  I   7.100  I   7.000
1897  I 14.543  I   4.640  I  .150  I  18.097  I   7.600  I   8.000
1898  I 12.354  I   5.260  I  .154  I  19.412  I   8.400  I   9.000
1899  I  6.536  I   6.090  I  .159  I  19.423  I   8.500  I  10.000
1900  I  5.004  I   6.600  I  .165  I  19.412  I   9.300  I  11.000
1901  I  4.133  I   7.480  I  .165  I  19.682  I   9.500  I  12.000
1902  I  3.668  I   8.170  I  .159  I  20.045  I   8.700  I  13.000
1903  I  3.922  I   8.680  I  .171  I  20.982  I   9.200  I  14.000
1904  I  5.378  I   9.240  I  .173  I  20.967  I   9.100  I  15.000
1905  I  4.276  I  10.240  I  .177  I  21.797  I   9.900  I  16.000
1906  I  1.728  I  11.080  I  .181  I  22.016  I  10.500  I  17.000
1907  I  2.764  I  11.600  I  .188  I  23.342  I  10.800  I  18.000
1908  I  7.962  I  11.440  I  .187  I  24.990  I  10.100  I  19.000
1909  I  5.106  I  12.680  I  .185  I  23.775  I  10.200  I  20.000
1910  I  5.857  I  13.340  I  .191  I  24.297  I  10.500  I  21.000
1911  I  6.719  I  14.120  I  .187  I  26.845  I  11.800  I  22.000
1912  I  4.637  I  15.130  I  .196  I  26.853  I  12.700  I  23.000
1913  I  4.321  I  15.370  I  .195  I  26.438  I  13.100  I  24.000
1914  I  7.919  I  16.390  I  .197  I  27.819  I  11.400  I  25.000
1915  I  8.528  I  17.590  I  .207  I  28.434  I  17.000  I  26.000
1916  I  5.100  I  20.850  I  .233  I  27.390  I  22.700  I  27.000
1917  I  4.617  I  24.370  I  .287  I  35.708  I  21.700  I  28.000
1918  I  1.372  I  26.730  I  .314  I  71.005  I  20.400  I  29.000
1919  I  1.375  I  31.010  I  .370  I  45.701  I  25.400  I  30.000
1920  I  5.157  I  34.800  I  .429  I  30.438  I  21.100  I  31.000
1921  I 11.715  I  32.850  I  .350  I  34.031  I  14.200  I  32.000
1922  I  6.728  I  33.720  I  .324  I  33.416  I  13.800  I  33.000
1923  I  2.415  I  36.600  I  .337  I  33.360  I  14.700  I  34.000
1924  I  4.951  I  38.580  I  .332  I  35.408  I  15.900  I  35.000
1925  I  3.217  I  42.050  I  .339  I  37.082  I  16.700  I  36.000
1926  I  1.756  I  43.680  I  .334  I  37.116  I  17.000  I  37.000
1927  I  3.275  I  44.730  I  .325  I  39.083  I  17.600  I  38.000
1928  I  4.208  I  46.420  I  .329  I  40.242  I  18.400  I  39.000
1929  I  3.246  I  46.600  I  .328  I  41.000  I  21.372  I  40.000
1930  I  8.903  I  45.730  I  .316  I  44.800  I  17.004  I  41.000
1931  I 15.653  I  42.690  I  .289  I  46.300  I  12.465  I  42.000
1932  I 22.981  I  36.050  I  .257  I  44.300  I   9.738  I  43.000
1933  I 20.901  I  32.220  I  .251  I  42.900  I   9.553  I  44.000
1934  I 16.197  I  34.360  I  .273  I  48.400  I  10.985  I  45.000
```

Table A.7 (continued)

	UN		M		P		G		X		T
1935	14.389		39.070		.279		49.600		11.834		46.000
1936	9.970		43.480		.280		57.900		12.506		47.000
1937	9.182		45.680		.293		55.800		16.039		48.000
1938	12.468		45.810		.286		60.700		15.359		49.000
1939	11.273		49.270		.284		63.000		16.184		50.000
1940	9.508		55.200		.291		65.300		18.581		51.000
1941	5.994		62.810		.312		97.800		19.536		52.000
1942	3.095		71.160		.343		191.600		14.565		53.000
1943	1.773		89.910		.361		271.300		12.730		54.000
1944	1.226		106.820		.370		300.400		14.862		55.000
1945	1.931		126.630		.379		265.400		19.524		56.000
1946	3.946		138.730		.439		93.100		34.425		57.000
1947	3.894		146.000		.496		75.700		40.755		58.000
1948	3.756		148.110		.530		84.700		33.031		59.000
1949	5.934		147.460		.525		96.800		31.060		60.000
1950	5.286		150.810		.536		98.100		26.880		61.000
1951	3.314		156.450		.571		133.700		34.505		62.000
1952	3.032		164.920		.579		159.800		32.975		63.000
1953	2.910		171.190		.588		170.100		30.602		64.000
1954	5.551		177.160		.595		156.000		31.410		65.000
1955	4.386		183.690		.608		152.300		34.519		66.000
1956	4.132		186.870		.628		153.500		39.815		67.000
1957	4.270		191.820		.649		161.200		43.276		68.000
1958	6.805		201.120		.660		169.900		36.642		69.000
1959	5.469		210.500		.676		170.600		36.684		70.000
1960	5.529		212.600		.687		172.800		42.063		71.000
1961	6.689		223.700		.693		182.900		43.123		72.000
1962	5.540		236.700		.706		193.200		45.043		73.000
1963	5.667		252.000		.716		197.600		47.715		74.000
1964	5.180		267.800		.728		202.600		53.323		75.000
1965	4.522		289.200		.744		209.800		55.266		76.000
1966	3.794		311.900		.758		229.700		58.098		77.000
1967	3.846		335.900		.791		248.500		59.829		78.000
1968	3.578		366.000		.825		260.200		63.481		79.000
1969	3.508		389.800		.868		257.400		66.247		80.000
1970	4.942		406.000		.914		251.100		71.848		81.000
1971	5.936		453.100		.960		250.100		71.660		82.000
1972	5.593		500.900		1.000		253.100		77.500		83.000
1973	4.853		549.100		1.057		253.500		103.700		84.000
1974	5.577		595.400		1.149		261.200		127.219		85.000
1975	8.455		641.300		1.256		266.700		123.374		86.000
1976	7.690		704.600		1.321		266.800		129.359		87.000
1977	7.038		779.700		1.398		272.300		131.091		88.000
1978	6.022		846.700		1.501		277.800		146.482		89.000
1979	5.794		914.400		1.628		281.800		172.819		90.000

b) The IAS-SYSTEM

The IAS-SYSTEM (for Inter-Active Simulation) is a special purpose statistical software package for the econometric analysis of time series data. As such, its main competitors are packages like TSP, TROLL or SHAZAM, which today are heavily used in empirical econometric work. There is also some overlap with well established general purpose statistical packages (SPSS, BMDP, etc.) in the areas of regression and time series analysis.

The need to develop the IAS-SYSTEM arose in the early 1970's when the Vienna Institute for Advanced Studies took charge of the Austrian part of the international LINK project (a research effort to assemble macroeconometric models of various nations into a comprehensive model of the world economy, headed by Nobel laureate Lawrence Klein), which involves the estimation and simulation of large scale econometric equation systems. At that time there was no suitable software compatible with the Institute's mainframe computer (a UNIVAC 1100), so the IAS-SYSTEM project was begun by the Institute's computer science and economics departments. A distinguishing feature of the IAS-SYSTEM has therefore from the beginning been a close collaboration between computer scientists, statisticians, and people who eventually use the system.

Thanks in part to a grant from the Austrian Forschungsförderungsfonds, resources devoted to the further development of the IAS-SYSTEM have since continually grown. Today the IAS-SYSTEM is among the most comprehensive econometric software packages on the market.

Data management

The IAS-SYSTEM has from the beginning been designed to be as interactive as possible in order to attain a maximum ease of the man-machine dialogue. There are currently about 1000 error messages, collected together on a separate message file, to guide users through the econometric modelling process.

The IAS-SYSTEM provides a permanent personal working area for every individual user to store and manipulate data items such as time series, equations or models. Input of data is either via terminal (format free) or through interfaces to external data banks. Special commands facilitate any updating, deleting or copying of data, and the FORMAT and TAB processors allow the self-tailored output of data ready for print. Users can also adjust the size of their data bases for efficient use of mass storage, or specify read and write keys to prevent unauthorized use. Additional help in data manipulation is provided by a built-in desk calculator (the CALC-command), various intrinsic functions, and the possibility to specify IF-THEN-ELSE constructs for data assignment. The system also allows for the storage and manipulation of text items such as entire run streams, which helps avoid unneccessary input activities in case of similar and recurring jobs.

The internal organization of the data follows a modified B^*-tree concept. This provides fast access to data items and is particularly convenient when simulating large scale econometric models, i.e. when up to several hundred equations have to be located in the data base.

All data are in the IAS-SYSTEM indexed by time.

The IAS-SYSTEM distinguishes annual, semi-annual, quarterly and monthly data. The type of periodicity and the aggregation mode (stock, flow, index number) are stored together with the data proper, so that for instance, when variables of different periodicity are entered into a regression, all neccessary aggregations can be performed automatically. To the best of our knowledge, this feature is not available elsewhere.

Seasonal adjustment of quarterly and monthly time series can also easily be done via the well known Census X-11 method, using an adapted version of the original source code of the U.S. Department of Commerce.

Econometric methods

The IAS-SYSTEM provides all standard estimation routines such as OLS, distributed lags, nonlinear least squares, or various IV- and systems-estimators. Recent additions also include Least-Absolute Deviations (LAD) estimation, various other robust estimators, plus time series routines such as the estimation of the order and the parameters of ARIMA processes, or spectral density estimation.

A major focus of the IAS-SYSTEM however is on testing rather than estimating econometric models. Most of the available procedures have been described in this monograph, so there is no need to go into detail here. Unlike many competing packages, the IAS-SYSTEM also computes prob-values for almost all test statistics and thus relieves the user of the sometimes arduous task of consulting statistical tables to evaluate the test results.

As its initial raison d'être, the IAS-SYSTEM also provides for the composition, simulation and forecasting of large scale econometric models, and has in this capacity for some years been used to forecast the Austrian economy. Models are solved with the Gauss-Seidel algorithm, augmented with a special acceleration algorithm to speed up convergence.

Another application area currently under rapid development concerns corporate planning and management consulting functions, such as linear and quadratic programming, to assist a growing number of users in the business community.

Hardware requirements and availability

The early versions of the IAS-SYSTEM (i.e. levels 1.X and 2.X) were written in Fortran IV and were tailor-made for UNIVAC machines. The current level 3.X of the

IAS-SYSTEM has been completely rewritten in Fortran 77 according to Standard ANSI-X3.9-1978, so as to be available on all machines with a Fortran 77 compiler. Modules that due to different interpretations of the standard (as happens with the OPEN, CLOSE or INQUIRE statements) might cause portability problems are isolated in a few subroutines, including some date and time routines, which have to be checked whenever the IAS-SYSTEM is transferred to a new machine.

At present there exist successful implementations on Univac, IBM, DEC, Siemens and CDC machines.

The IAS-SYSTEM comprises about 600 subroutines (with 120000 lines of source code, including comment lines) and, when using simple overlay structures, requires about 150K words main storage. Due to this size, it is not yet available on microcomputers.

However, the system is fully parameterized in the sense that the maximum length of time series or the maximum number of variables in a regression can be either reduced to save main storage or enlarged to satisfy particular user needs.

The IAS-SYSTEM has initially been designed for home use only, but has since 1977 been made available to outside users on a licensing basis. At present, there are about 50 customers in 12 countries worldwide. License fees vary across institutions (in particular, there are special rates for universities), and in addition depend upon whether the source code or only the object code is desired.

Academic institutions can further reduce the annual licence fee by offering an introduction to the IAS-SYSTEM, or by using the IAS-SYSTEM in other courses (i.e. on policy simulation or empirical econometrics). The fees then reduce by 50% for the corresponding license period, except for the first.

Further information is available in *Sonnberger* (1985), *Sonnberger* et.al (1986a,b), and from the IAS-SYSTEM project group, Institute for Advanced Studies, Stumpergasse 56, A-1060 Vienna, Austria.

An example

The following example - a simple regression - is meant to demonstrate the IAS-SYSTEM syntax and command language and does not intend to provide an overview of the available statistical methods.

Suppose C_t and Y_t (t = 1960,...,1984) are annual aggregate consumption and income data for some geographical entity. A favourite pastime of applied economists is to specify a model

$$C_t = b_1 + b_2 Y_t + u_t$$

and to estimate b_1 and b_2 from the data. In the IAS-SYSTEM, this is done with the following series of commands (assuming that the relevant data have not yet been stored and that this is the user's first contact with the system):

*db,c name

This command assigns a data base to the run. "Name" can be any legal name for a permanent file on the underlying mainframe and will subsequently identify the data base. The option "c" tells the system that data base "name" does not yet exist and has to be newly created. This is one of a variety of options that can be attached to *db and other IAS-SYSTEM commands. Options are separated from the commands proper by commas, and from the rest of the command by blanks. Note also that all IAS-commands begin with an asterisk.

*f,c name

This assigns a (previously nonexistent) file "name" to the job, where all data are to be stored. Subdividing data bases into files is mainly for ease of data manipulation.

*time,d 1960:1984

This tells the system to expect annual data beginning in 1960 and ending in 1984. Quarterly data are indicated by e.g. *time,d 1968q1:1978q4, and the system expects monthly data after e.g. *time,d 1979m3:1983m12.

The data themselves are now entered via

*ser,i c

where "i" means "input" and "c" specifies the name of the time series (an optional header such as "consumption expenditures" can also be stored). Next, the user types in the consumption data, either one in a row or separated by blanks, and repeats the same procedure for the income data.

*time,e 1960:1984

This tells the system that for any subsequent estimation command, the respective data will cover the range 1960-1984.

*OLS

C,Y

*

This finally performs the desired Ordinary Least Squares (OLS) regression of C on Y, including an output that is mostly self-explanatory. If satisfied with the result, the user types

*exit ,

which terminates the IAS-SYSTEM and returns control to the operating system of the underlying mainframe. The data will then be saved and will be immediately available in future runs.

A distinguishing feature of the IAS-SYSTEM is the command *test, which can be called after any single equation estimation routine, and whose theoretical background has been the subject of the present monograph.

The command *test remains active until terminated via * or via another IAS-SYSTEM command. Specific tests are called via subcommands, which are characterized by ** (e.g. **haus).

To ensure identical test results for repeated calls of the same subcommand within one call of *test, both data and results of the last regression are set aside on a separate working area, and cannot be destroyed by individual tests. The *test command is thus an endless loop allowing an arbitrary number of tests, without having to reestimate the equation for each individual test.

REFERENCES

ABRAHAMSE, A.P.J., and Louter, A.S. (1971): "On a New Test for Autocorrelation in Least Squares Regression", *Biometrika* 58, 53-60.

ALDRICH, J. (1978): "An Alternative Derivation of Durbin's h Statistic", *Econometrica* 46, 1493-1494.

ALLEN, Stuart D., (1982): "Klein's Price Variability Terms in the U.S. Demand for Money", *Journal of Money, Credit and Banking* 14, 525-530.

ALT, R. (1985): "Neuere Untersuchungen zum CUSUM Test", *Research Memorandum* No. 225, Institute for Advanced Studies, Vienna.

ANDERSON, G.J., and Mizon, G.E. (1983): "Parameter Constancy Tests: Old and New", *Discussion Paper No. 8325*, Dep. of Economics, University of Southhampton.

ALT, R., and Krämer, W. (1986): "A Modification of the CUSUM Test", Research Memorandum No. 229, Institute for Advanced Studies, Vienna.

ANDERSON, T.W. (1948): "On the Theory of Testing Serial Correlation", *Skandinavisk Aktuarietidskrift* 31, 88-116.

ANDERSON, T.W. (1975): Discussion of the paper by Brown, Durbin and Evans, *JRSS* B37, p.175.

ANEURYN EVANS, G., and Deaton, A.S. (1980): "Testing Linear Versus Logarithmic Regression Models", *Review of Economic Studies* 47, 275-291.

ANSCOMBE, F.J. (1961): "Examination of Residuals", *Proceedings of the Fourth Berkeley Symposium on Mathematical Statistics and Probability*, 4, 1-36.

ATHANS, M. (1974): "The Importance of Kalman Filtering Methods for Economic Systems", *Annals of Economic and Social Measurement* 3, 49-64.

BARTELS, R., Bornholt, G., and Hanslow, K. (1982): "The Polynomial Trend Model with Autocorrelated Residuals", *Comm. Stat. (Theory and Methods)* 11, 1393-1402.

BAUER, P. and Hackl, P. (1978): "The Use of MOSUM's for Quality Control", *Technometrics* 20, 431-436.

BERA, A.K., and Jarque, C.M. (1982): "Model Specification Tests: A Simultaneous Approach", *Journal of Econometrics* 20, 59-82.

BERENBLUT, I.I., and Webb, G.I. (1973): "A New Test for Autocorrelated Errors in the Linear Regression Model, *JRSS* B35, 33-50.

BERNDT, E.K., Hall, B.H., Hall, R.E., and Hausman, J.A. (1974): "Estimation and Inference in Nonlinear Structural Models", *Annals of Ec. Soc. Measurement* 3, 653-666.

BIERENS, H.J. (1982): "Consistent Model Specification Tests", *Journal of Econometrics* 20, 105-134.

BORDO, Michael D., and Choudhri, Ehsan U. (1982): "Currency Substitution and Demand for Money", *Journal of Money, Credit and Banking* 14, 48-57.

BELSLEY, D.A. (1974): "Estimators of System of Simultaneous Equations and Computational Specification of Gremlin", *Annals of Economic and Social Measurement* 3/4, 551-614.

BILLINGSLEY, P. (1968): *Convergence of Probability Measures*, New York (Wiley).

BREIMAN, L. (1968): *Probability*, Reading (Addison-Wesley).

BREUSCH, T.S. (1978): "Testing for Autocorrelation in Dynamic Linear Models", *Australian Economic Papers* 17, 334-355.

BREUSCH, T.S., and Godfrey, L.G. (1984): "Implementation, Power and Interpretation of the Differencing Test", *University of Southhampton Discussion Paper* No. 8412.

BREUSCH, T.S., and Godfrey, L.G. (1985): "Data Transformation Tests," Paper given at the 5th World Congress of the Econometric Society, Cambridge, MA.

BREUSCH, T.S., and Pagan, A.R. (1979): "A Simple Test for Heteroscedasticity and Random Coefficient Variation", *Econometrica* 47, 1287-1294.

BREUSCH, T.S., and Pagan, A.R. (1980): "The Lagrange Multiplier Test and its Applications to Model Specification in Econometrics", *Review of Economic Studies* 47, 239-253.

BREUSCH, T.S., and Pagan, A.R. (1984): "Model Evaluation by Differencing Transformations", Mimeo, Australian National University.

BROWN, R.L., Durbin, J., and Evans, J.M. (1975): "Techniques for Testing the Constancy of Regression Relationships over Time", *JRSS* B 37, 149-192.

BUSE, A. (1982): "The Likelihood Ratio, Wald, and Lagrange Multiplier Tests: An Expository Note", *The American Statistician* 36, 153-157.

den BUTTER, F.A.G., and Verbon, H.A.A. (1982): "The Specification Problem in Regression Analysis", *International Statistical Review* 50, 267-283.

CHIPMAN, J.S. (1965): "The Problem of Testing for Serial Correlation in Regression Analysis: The Story of a Dilemma", Technical Report No.4, Dep. of Economics, The University of Minnesota.

CHOW, C. (1960): "Tests for Equality between Sets of Coefficients in Two Linear Regressions", *Econometrica* 38, 591-605.

COOK, Dennis R., and Weisberg, Sanford (1982): *Residuals and Influence in Regression*, New York - London (Chapman & Hall).

COOK, Timothy Q., and Hendershott, Patricia H. (1978): "The Impact of Taxes, Risk and Relative Security Supplies on Interest Rate Differentials", *The Journal of Finance* 33, 1173-1186.

COOLEY, Th.F., and Prescott, E.C. (1976): "Estimation in the Presence of Stochastic Parameter Variation", *Econometrica* 44, 167-184.

DAVID, H.A. (1981): *Order Statistics* (2nd Ed.), New York (Wiley).

DAVIDSON, R., and MacKinnon, J.G. (1984): "A Simplified Version of a Differencing Specification Test", Queen's University Institute for Economic Research, *Discussion Paper* No. 548.

DAVIDSON, R., Godfrey, L., and MacKinnon, J.G. (1986): "A Simplified Version of the Differencing Test", to appear in *International Economic Review*.

DENT, W.T., and Styan, G.P.H. (1978): "Uncorrelated Residuals from Linear Models", *Journal of Econometrics* 7, 211-225.

DHRYMES, Ph.(1978): *Introductory Econometrics*, New York (Springer).

DUBBELMAN, C. (1972): "A Priori Fixed Covariance Matrices of Disturbance Estimators", *European Economic Review* 3, 413-436.

DUBBELMAN, C. (1978): *Disturbances in the Linear Model: Estimation and Hypothesis Testing*, The Hague (M.Nijhoff).

DUBBELMAN, C., Louter, A.S., and Abrahamse, A.P.J. (1978): "On Typical Characteristics of Economic Time Series and the Relative Qualities of Five Autocorrelation Tests", *Journal of Econometrics* 8, 295-306.

DUFOUR, J.M. (1980): "Dummy Variables and Predictive Tests for Structural Change", *Economics Letters* 6, 241-247.

DUFOUR, J.M. (1982): "Recursive Stability Analysis of Linear Regression Relationships", *Journal of Econometrics* 19, 31-76.

DURBIN, J. (1954): "Errors in Variables", *Review of the International Statistical Institute* 22, 23-32.

DURBIN, J. (1969): "Test for Serial Correlation in Regression based on the Periodogram of Least Squares Residuals", *Biometrika* 56, 1-15.

DURBIN, J. (1970): "Testing for Serial Correlation in Least Squares Regression when Some of the Regressors are Lagged Dependent Variables, *Econometrica* 38, 410.

DURBIN, J. (1971): "Boundary-Crossing Probabilities for the Brownian Motion and Techniques for Computing the Power of the Kolmogorov-Smirnov Test", *Journal of Applied Probability* 8, 431-453.

DURBIN, J., and Watson, G.S. (1950): "Testing for Serial Correlation in Least Squares Regression I", *Biometrika* 37, 409-428.

DURBIN, J., and Watson, G.S. (1951): "Testing for Serial Correlation in Least Squares Regression II", *Biometrika* 38, 159-178.

DURBIN, J., and Watson, G.S. (1971): "Testing for Serial Correlation in Least Squares Regression III", *Biometrika* 58, 1-19.

ENGLE, R.F. (1984): "Wald, Likelihood Ratio, and Lagrange Multiplier Tests in Econometrics", in: Griliches, Z., and Intriligator, M. (eds.): *Handbook of Econometrics*, Amsterdam (North-Holland).

EPPS, T.W., and Epps, M.L. (1977): "The Robustness of Some Standard Tests for Autocorrelation and Heteroscedasticity when Both Problems are Present", *Econometrica* 45, 745-753.

FAIR, R.C, and Jaffee, D. (1972): "Methods of Estimation for Markets in Disequilibrium", *Econometrica* 40, 497-514.

FAREBROTHER, R.W. (1979): "A Grouping Test for Misspecification", *Econometrica* 47, 209-210.

FAREBROTHER, R.W. (1980): "The Durbin-Watson Test for Serial Correlation when there is no Intercept in the Regression", *Econometrica* 48, 1553-1563.

FARLEY, J.U., and Hinich, M. (1970): "Testing for a Shifting Slope Coefficient in a Linear Model", JASA 65, 1320-1329.

FARLEY, J.U., Hinich, M., and McGuire, T.W. (1975): "Some Comparison of Tests for a Shift in the Slopes of a Multivariate Linear Time Series Model", *Journal of Econometrics* 3, 297-318.

FRENKEL, J.A. (1980a): "Exchange Rates, Prices and Money: Lessons from the 1920's", *American Economic Review* 70, 235-245.

FRENKEL, J.A. (1980b): "The Forward Exchange Rate, Expectations and the Demand for Money - The German Hyperinflation Reply", *American Economic Review* 70, 771-775.

FRENKEL, J.A. (1981): "Flexible Exchange Rates, Prices and the Role of 'News': Lessons from the 1970's", *Journal of Political Economy* 89, 665-705.

FRIEDMAN, R. (1977): "Trendbereinigung mit ersten Differenzen - Eine Klarstellung", *Statistische Hefte* 18, 203-208.

GÄNSSLER, P. and Stute, W. (1977): *Wahrscheinlichkeitstheorie*, Berlin (Springer).

GARBADE, K. (1977): "Two Methods for Examining the Stability of Regression Coefficients", JASA 72, 54-63.

GASTWIRTH, J.L., and Owens, M.E.B. (1977): "On Classical Tests of Normality", *Biometrika* 64, 135-139.

GILES, D.E.A., and King, M. (1978): "Fourth Order Autocorrelation: Further Significance Points for the Wallis Test", *Journal of Econometrics* 8, 255-260.

GLEJSER, H. (1969): "A New Test for Heteroskedasticity", JASA 64, 316-323.

GODFREY, L. (1978a): "Testing Against General Autoregressive and Moving Error Models when the Regressors Include Lagged Dependent Variables", *Econometrica* 46, 1293-1302.

GODFREY, L. (1978b): "Testing for Higher Order Serial Correlation in Regression Equations when the Regressors Include Lagged Endogeneous Variables", *Econometrica* 46, 1303-1310.

GODFREY, L. (1978c): "Testing for Multiplicative Heteroscedasticity", *Journal of Econometrics* 8, 227-236.

GODFREY, L.G. (1984): "On the Interpretation and Implementation of the Differencing Test", University of York (Mimeo).

GODFREY, L., and Tremayne, A.R. (1978): "Testing for Fourth Order Autocorrelation in Dynamic Quarterly Regression Equations: Asymptotic Tests and Monte Carlo Evidence", unpublished manuscript, University of York.

GODFREY, L., and Wickens, M.R. (1981): "Testing Linear and Log-Linear Regressions for Functional Form", *Review of Economic Studies* 48, 487-496.

GOLDBERG, Lawrence G., and Saunders, A. (1981): "The Growth of Organizational Forms of Foreign Banks in the U.S.", *Journal of Money, Credit and Banking* 13, 365-374.

GOLDFELD, S.M., and Quandt, R.E. (1965): "Some Tests for Homoskedasticity", JASA 60, 539-547.

GREENLEES, J.S., and Zieschang, K.D. (1984): "Grouping Tests for Misspecification: An Application to Housing Demand", *Journal of Business and Economic Statistics* 2, 159-169.

HABIBAGHI, H., and Pratschke, J.L. (1972): "A Comparison of the Power of the von Neumann Ratio, Durbin-Watson and Geary Tests", *Review of Economics and Statistics* 54, 179-185.

HACKL, P. (1980): *Testing the Constancy of Regression Models over Time*, Göttingen (Vandenhoek & Ruprecht).

HACKL, P., and Westlund, A. (1985): "Statistical Analysis of the Structural Change: Annotated Bibliography", *Collaborative Paper* No. CP-85-31, International Institute for Applied System Analysis, Laxenburg.

HALL, A.R. (1983): "The Information Matrix Test for the Linear Model", paper given at the Econometric Society Meeting, Pisa.

HAN, Chien-Pai (1985): "NFD: Noncentral F-distribution", *The American Statistician* 39, p.211.

HARRISON, M.J. (1972): "On Testing for Serial Correlation in Regression when the Bounds Test is Inconclusive", *Economic and Social Review* 4, 41-57.

HARRISON, M.J., and McCabe, B.P.M. (1979), "A Test for Heteroscedasticity based on Ordinary Least Squares Residuals", JASA 74, 494-499.

HARVEY, A. (1975): Contribution to the discussion of the paper by Brown, Durbin and Evans, JRSS B 37, 179-180.

HARVEY, A. (1976): "Estimating Regression Models with Multiplicative Heteroskedasticity", *Econometrica* 44, 461-465.

HARVEY, A. (1981): *The Econometric Analysis of Time Series*, Oxford (Philip Allan).

HARVEY, A., and Collier, P. (1977): "Testing for Functional Misspecification in Regression Analysis", *Journal of Econometrics* 6, 103-119.

HARVEY, A., and Phillips, G.D.A. (1980): "Testing for Serial Correlation in Simultaneous Equation Models", *Econometrica* 48, 747-759.

HAUSMAN, J.A. (1978): "Specification Tests in Econometrics", *Econometrica* 46, 1251-1271.

HAUSMAN, J.A., and Taylor, W.E. (1980): "Comparing Specification Tests and Classical Tests", Mimeo, MIT.

HAUSMAN, J.A. and Taylor, W.E. (1981): "A Generalized Specification Test", *Economics Letters* 8, 239-245.

HAWKINS, D.M. (1977): "Point Estimation of the Parameters Piecewise Regression Models", *Applied Statistics* 25, 51-57.

HENDRY, D.F. (1980): "Econometrics: Alchemy or Science?", *Economica* 47, 387-406.

HETZEL, Robert L. (1981): "The Federal Reseve System and Control of the Money Supply in the 1970's", *Journal of Money, Credit and Banking* 13, 31-43.

HILDRETH, C., and Houck, P. (1968): "Some Estimators of a Linear Model with Random Coefficients", JASA 63, 584-595.

HINKLEY, D.V., Chapman, P., and Ringer, G. (1980): "Change Point Problem", Technical Report No.382, University of Minnesota, School of Statistics.

HOAGLIN, D.C., and Welsch, R.E. (1978): "The Hat Matrix in Regression and ANOVA", *The American Statistician* 32, 17-22.

HOLLY, A. (1982): "A Remark on Hausman's Specification Test", *Econometrica* 50, 749-759.

HOLM, S. (1979): "A Simple Sequentially Rejection Multiple Test Procedure", *Scandinavian Journal of Statistics* 6, 65-70.

HONDA, Y. (1982): "On Tests of Equality between Sets of Coefficients in Two Linear Regressions when Disturbance Variances are Unequal", *The Manchester School* 50, 116-125.

HUANG, C.J., and Bolch, B.W. (1974): "On the Testing of Regression Disturbances for Normality", JASA 69, 330-335.

HUXLEY, J.S. (1972): *Problems of Relative Growth* (2nd Ed.), New York (Dover).

IMHOF, J.P. (1961): "Computing the Distribution of Quadratic Forms in Normal Variables", *Biometrika* 48, 419-426.

JARQUE, C.M., and Bera, A.K. (1980): "Efficient Tests for Normality, Homoskedasticity and Serial Independence of Regression Residuals", *Economics Letters* 6, 255-259.

JAYATISSA, W.A. (1977): "Tests of Equality Between Sets of Coefficients in Two Linear Regressions when Disturbance Variances are Unequal", *Econometrica* 45, 1291-1292.

JOHNSTON, J. (1984): *Econometric Methods* (3rd ed.), New York (McGraw-Hill).

JUDGE, G.G., Griffiths, W.E., Hill, R.C., and Lee, T.Ch. (1980): *The Theory and Practice of Econometrics*, New York (Wiley).

KENDALL, M.G., and Stuart, A. (1969): *The Advanced Theory of Statistics*, Vol. I, 3rd. Ed., London (Griffin).

KING, M.L. (1981a): "The Durbin-Watson Test for Serial Correlation: Bounds for Regressions with Trend and/or Seasonal Dummy Variables", *Econometrica* 49, 1571-1582.

KING, M.L. (1981b): "The Durbin-Watson Test and Regressions without an Intercept", *Australian Economic Papers* 20, 161-170.

KING, M.L. (1982): "A Locally Optimal Bounds Test for Autoregressive Disturbances", Paper presented at the European Meeting of the Econometric Society, Dublin.

KING, M.L. (1983a): "Testing for Autocorrelation in Linear Regression Models: A Survey", Working paper No. 7/83, Dep. of Econometrics, Monash University (to appear in King, M.L. and Giles, D.E.A. (eds.): *Specification Analysis in the Linear Model: Essays in Honour of Donald Cochrane*).

KING, M.L. (1983b): "The Durbin-Watson Test for Serial Correlation: Bounds for Regressions Using Monthly Data", *Journal of Econometrics* 21, 357-366.

KING, M.L. (1985): "A Point Optimal Test for Autoregressive Disturbances", *Journal of Econometrics* 27, 21-37.

KLEIN, Benjamin (1977): "The Demand for Quality-adjusted Cash Balances: Price Uncertainty in the U.S. Demand for Money Function", *Journal of Political Economy* 85, 691-715.

KOENKER, R. (1981): "A Note on Studentizing a Test for Heteroskedasticity", *Journal of Econometrics* 17, 107-112.

KOERTS, J., and Abrahamse, A.P.J. (1969): *On the Theory and Application of the General Linear Model*, Rotterdam (Rotterdam University Press).

KONTRUS, K. (1984): "Monte Carlo Simulationen zu zwei Tests auf Konstanz der Parameter in Linearen Modellen", Diplomarbeit, Technische Universität Wien.

KONTRUS, K., and Ploberger, W. (1984): "A Simulation of Some Tests for Parameter Constancy in Linear Models", paper given at the Econometric Society European Meeting, Madrid.

KRAMER, G. (1971): "On the Durbin-Watson Bounds Test in the Case of Regression Through the Origin", *Jahrbücher für Nationalökonomie und Statistik* 185, 345-358.

KRÄMER, W. (1984a): "Some Consequences of Trend for Simultaneous Equation Estimation", *Economics Letters* 14, 23-30.

KRÄMER, W. (1984b): "Ordinary Least Squares Estimation of the Functional Errors-in-Variables Model with Trended Data: Some Monte Carlo Evidence", *Communications in Statistics B* 13, 655-666.

KRÄMER, W. (1985a): "Ordinary Least Squares Estimation of Simultaneous Equation Systems with Trended Data: Further Results", *Communication in Statistics - Theory and Methods* 14, 1997-2005.

KRÄMER, W. (1985b): "The Power of the Durbin-Watson Test for Regression without an Intercept", *Journal of Econometrics* 28, 363-370.

KRÄMER, W. (1985c): "A Hausman Test with Trending Data", *Economics Letters* 19, 323-325.

KRÄMER, W. (1985d): "Asymptotische Verteilung einiger Schätzverfahren bei Trend und Fehlern in den Variablen", *Metrika* 32, 151-162.

KRÄMER, W., Sonnberger, H., Maurer, J., and Havlik, P. (1984): "Diagnostic Checking in Practice", *Review of Economics and Statistics* 12, 118-123.

KRÄMER, W., Ploberger, W., and Alt, R. (1985): "Testing for Structural Change in Dynamic Models", unpublished manuscript.

KRÄMER, W. and Sonnberger, H. (1986): "Computational Pitfalls of the Hausman Test", to appear in *Journal of Economic Dynamics and Control*.

L'ESPERANCE, W.L., and Taylor, D. (1975): "The Power of Four Tests of Autocorrelation in the Linear Regression Model", *Journal of Econometrics* 3, 1-21.

LOCKE, C., and Spurrier, J.D. (1977): "The Use of U-Statistics for Testing Normality Against Alternatives with Both Tails Heavy or Both Tails Light", *Biometrika* 64, 638-640.

MACKINNON, J.G. (1983): "Model Specification Tests Against Non-Nested Alternatives", *Econometric Reviews* 2, 85-110.

MADDALA, G.S. (1977): *Econometrics*, New York (McGraw-Hill).

MARDIA, K.V. (1980): "Tests for Univariate and Multivariate Normality", in: *Handbook of Statistics*, Vol. I, Amsterdam (North-Holland), 279-320.

MAURER, J., Sonnenberger, H., Havlik, P., and Kraemer, W. (1983): "The IAS-SYSTEM Level 3.3, User Reference Manual, Part Three:Testing", Institutsarbeit No.196, Institute for Advanced Studies.

McCABE, B.P.M. , and Harrison, M.J. (1980): "Testing the Constancy of Regression Relationships over Time Using Least Squares Residuals", JRSS C.29, 142-148.

MILES, Marc (1978): "Currency Substitution, Flexible Exchange Rates, and Monetary Independence", *American Economic Review* 68, 428-436.

MILLER, R.G. (1966): *Simultaneous Statistical Inference*, New York.

MILLER, R.G. (1977): "Developments in Multiple Comparisons", JASA 72, 779-788.

MORGENSTERN, D. (1980): "Berechnung des maximalen Signifikanzniveaus des Testes: Lehne H ab, wenn k unter n gegebenen Tests zur Ablehnung führen", *Metrika* 27, 285-286.

NAKAMURA, A., and Nakamura, M. (1981): "On the Relationships among Several Specification Error Tests presented by Durbin, Wu and Hausman", *Econometrica* 49, 1583-1588.

NERLOVE, M., and Wallis, K.F. (1966): "Use of the Durbin-Watson Statistic in Inappropriate Situations", *Econometrica* 34, 235-238.

NEUWIRTH, E. (1982): "Parametric Derivations in Linear Models", in: Grossman, W. et.al. (eds.): *Probability and Statistical Inference*, Dordrecht (Reidel).

NICHOLLS, D.F., and Pagan, A.R. (1983): "Heteroskedasticity in Models with Lagged Dependent Variables", *Econometrica* 52, 1233-1242.

NICHOLS, Donald A. (1983): "Macroeconomic Determinants of Wage Adjustments in White Collar Occupations", *Review of Economics and Statistics* 65, 203-213.

PAGAN, A.R. (1982): "Commentary on Model Selection", Paper given at the Meeting of Australasian Econometricians, Melbourne.

PAGAN, A.R. (1984): "Model Evaluation by Variable Addition", in: Hendry, D.F., and Wallis, K.F. (eds.): *Econometrics and Quantitive Economics*, Oxford.

PAGAN, A.R., and Hall, A.D. (1983): "Diagnostic Tests as Residual Analysis", *Econometric Reviews* 2, 159-218.

PAGAN, A.R., and Nicholls, D.F. (1984): "Estimating Predictions, Prediction Errors and their Standard Deviations Using Constructed Variables", *Journal of Econometrics* 24, 293-310.

PEARSON, E.S., D'Agostino, R.B.D., and Bowman, K.O. (1977): "Tests for Departure from Normality: Comparison of Powers", *Biometrika* 64, 231-246.

PEARSON, E.S., and Hartley, H.O. (1966): *Biometrika Tables for Statisticians*, Vol.2, Cambridge (Cambridge University Press).

PESARAN, H.M., Smith, R.P., and Yeo, J.S. (1985): "Testing for Structural Stability and Predictive Failure: A Review", *The Manchester School* 53, 281-295.

PHILLIPS, G.D.A. (1984): "Some Recent Developments in the Econometric Test Methodology", Research Memorandum No. 205, Institute for Advanced Studies, Vienna.

PHILLIPS, G.D.A., and Harvey, A. (1974): "A Simple Test for Serial Correlation in Regression Analysis", JASA 69, 935-939.

PHILLIPS, G.D.A., and McCabe, B.P. (1983): "The Independence of Tests for Structural Change in Regression Models", *Economics Letters* 12, 283-287.

PLOBERGER, W. (1983): "Testing the Constancy of Parameters in Linear Models", Paper presented at the European Meeting of the Econometric Society, Pisa.

PLOBERGER, W., and Krämer, W. (1986a): "On Studentizing a Test for Structural Change", *Economics Letters* 20, 341-344.

PLOBERGER, W., and Krämer, W. (1986b): "The Local Power of the CUSUM and CUSUM of Squares Test", unpublished manuscript.

PLOBERGER, W., Kontrus, K., and Krämer, W. (1986): "A New Test for Structural Stability in the Linear Regression Model", unpublished manuscript.

PLOSSER, Ch.I., Schwert, G.W., and White, H. (1982): "Differencing as a Test of Specification", *International Economic Review* 23, 535-552.

POIRIER, D.J. (1973): "Piecewise Regression Using Cubic Splines", JASA 68, 515-524.

POIRIER, D.J. (1976): *The Econometrics of Structural Change*, Amsterdam (North-Holland).

POLINSKY, A.M. (1977): "The Demand for Housing: A Study in Specification and Grouping", *Econometrica* 45, 447-461.

PYKE, R. (1959): "The Supremum and the Infimum of the Poison Process", *Annals of Mathematical Statistics* 30, 569-576.

QUANDT, R.E. (1958): "The Estimation of the Parameters of a Linear Regression Obeying two Separate Regimes", JASA 53, 873-880.

QUANDT, R.E. (1960): "Tests of the Hypothesis that a Linear Regression System Obeys two Separate Regimes", JASA 55, 324-330.

RAMSEY, J.B. (1969): "Tests for Specification Error in Classical Linear Least Squares Regression Analysis", JRSS B31, 350-371.

RAMSEY, J.B. (1974): "Classical Model Selection through Specification Error Tests", in Zarembka, P.: *Frontiers in Econometrics*, New York (Academic Press), 13-47.

RAMSEY, J.B., and Schmidt, P. (1976): "Some Further Results on the Use of OLS and BLUS Residuals in Specification Error Tests", JASA 71, 389-390.

RAO, C.R. and Mitra, S.K. (1971): *Generalized Inverse of Matrices and its Applications*, New York (Wiley).

REA, John D. (1983): "The Explanatory Power of Alternative Theories of Inflation and Unemployment, 1895-1979", *Review of Economics and Statistics* 65, 183-195.

ROY, S.N. (1953): "On a Heuristic Method of Test Construction and Its Uses in Multivariate Analysis", *Annals of Mathematical Statistics* 24, 220-239.

RÜGER, B. (1978): "Das maximale Signifikanzniveau des Tests: Lehne H_0 ab, wenn k unter n gegebenen Tests zur Ablehnung fuehren", *Metrika* 25, 171-178.

SALKEVER, D.S. (1976): "The Use of Dummy Variables to Compute Predictions, Prediction Errors and Confidence Intervals", *Journal of Econometrics* 4, 393-397.

SARGAN, D. (1975): Contribution to the discussion of a paper by D.Hendry, in *Modelling the Economy*, edited by G.A.Renton, London.

SAVIN, N.E. (1984): "Multiple Hypothesis Testing", in: *Handbook of Econometrics*, Vol.2, (edited by Z.Griliches and M.D.Intriligator), Amsterdam (North Holland).

SAVIN, N.E., and White, K.J. (1977): "The Durbin-Watson Test for Serial Correlation with Extreme Sample Sizes or Many Regressors", *Econometrica* 45, 1989-1996.

SCHMIDT, P. (1972): "A Generalization of the Durbin-Watson Test", *Australian Economic Papers* 11, 203-209.

SCHMIDT, P., and Guilkey, D.K. (1975): "Some Further Evidence on the Power of the Durbin-Watson and Geary Tests", *Review of Economics and Statistics* 57, 379-382.

SCHMIDT, P., and Sickles, R. (1977): "Some Further Evidence on the Chow Test under Heteroskedasticity", *Econometrica* 45, 1293-1298.

SCHWAGER, S.J. (1984): "Bonferroni Sometimes Looses", *The American Statistician* 38, 192-197.

SCHWEDER, T. (1976): "Some 'Optimal' Methods to Detect Structural Shift or Outliers in Regression", JASA 71, 491-501.

SEN, P.K. (1980): "Asymptotic Theory of Some Tests for a Possible Change in the Regression Slope Occuring at an Unknown Time Point", *Zeitschrift für Wahrscheinlichkeitstheorie und verwandte Gebiete* 52, 203-218.

SERFLING, A.(1980): *Approximation Theorems of Mathematical SHAPIRO, S.S., and Wilk, M.B. Statistics*, New York (Wiley).

(1965): "An Analysis of Variance Test for Normality (Complete Samples)", *Biometrika* 52, 591-611.

SMITH, V.K. (1977): "A Note on the Power of the Cumulated Periodogram Test for Autocorrelation", *European Economic Review* 9, 373-377.

SONNEMANN, E. (1982): "Tests zum multiplen Niveau α", in: *Simultane Hypothesenprüfungen* (edited by U.Ferner), Basel.

SONNBERGER, H. (1985): "An Introduction to the IAS-System", *Computational Statistics and Data Analysis* 2, 323-328.

SONNBERGER, H., Havlik, P., Krämer, W., and Maurer, J. (1983): "Testing for Autocorrelation and Model Specification with the IAS-SYSTEM", Research Memorandum No.186, Institute for Advanced Studies, Vienna.

SONNBERGER, H., Krämer, W., Schraick, W., Reschenhofer, E., Wasilewski, Z. and Zeisel, H. (1985): "IAS-System Level IAS-3.6 User Reference Manual - Part Two: Econometric Methods", *Institutsarbeit* Nr.234, Institute for Advanced Studies, Vienna.

SONNBERGER, H., Rodler, K., Plasser, K., Philipp, W., Schraick, W., Zolles, K., Scheiber, J. and Zeisel, H. (1985): "IAS-SYSTEM Level IAS-3.6 User Reference Manual - Part One: General", *Institutsarbeit* Nr.232, Institute for Advanced Studies, Vienna.

SPIEGELHALTER, D.J. (1977): "A Test for Normality Against Symmetric Alternatives", *Biometrika* 64, 415-209.

SPENCER, D.E., and Berk, K.N. (1981): "A Limited Information Specification Test", *Econometrica* 49, 1079-1085.

SPENCER, D.E., and Berk, K.N. (1982): "Erratum", *Econometrica* 50, 1087.

THEIL, H. (1965): "The Analysis of Disturbances in Regression Analysis", JASA 60, 1067-1079.

THEIL, H. (1971): *Principles of Econometrics*, New York (Wiley).

THURSBY, J.G. (1979): "Alternative Specification Error Tests: A Comparative Study", JASA 74, 222-225.

THURSBY, J.G. (1981): "A Test Strategy for Discriminating between Autocorrelation and Misspecification in Regression Analysis", *Review of Economics and Statistics* 63, 117-123.

THURSBY, J.G. (1982): "Misspecification, Heteroscedasticity, and the Chow and Goldfeld-Quandt Tests", *Review of Economics and Statistics* 64, 314-321.

THURSBY, J.G. and Schmidt, P. (1977): "Some Properties of Tests for Specification Error in a Linear Regression Model", JASA 72, 635-641.

TILLMAN, J.A. (1975): "The Power of the Durbin-Watson Test", *Econometrica* 43, 959-974.

TOYODA, T. (1974): "Use of the Chow Test under Heteroskedasticity", *Econometrica* 42, 601-608.

UTTS, Jessica M. (1982): "The Rainbow Test for Lack of Fit in Regression", *Communications in Statistics-Theory and Methods* 11, 1801-1815.

WALLIS, K.F. (1972): "Testing for Fourth-Order Autocorrelation in Quarterly Regression Equations", *Econometrica* 40, 617-636.

WATT, P.A. (1979): "Tests of Equality Between Sets of Coefficients in Two Linear Regressions when Disturbance Variances are Unequal: Some Small Sample Properties", *The Manchester School* 47, 391-396.

WEISBERG, S.(1980): *Applied Linear Regression*, New York (Wiley).

WEISBERG, S. (1980a): Comment on White and MacDonald: Some Large Sample Tests for Nonnormality in the Linear Regression Model, JASA 75, 28-31.

WHITE, H. (1980): "A WHITE, H. (1980): "A Heteroskedasticity-Consistent Covariance Matrix Estimator and a Direct Test for Heteroskedasticity", *Econometrica* 48, 817-838.

WHITE, H. (1980a): "Using Least Squares to Approximate Unknown Rgression Functions", *International Economic Review* 21, 149-170.

WHITE, H. (1982): "Maximum Likelihood Estimation of Misspecified Models", *Econometrica* 50, 1-25.

WHITE, H., and MacDonald, G. (1980): "Some Large-Sample Tests for Non-Normality in the Linear Regression Model", JASA 75, 16-28.

WOGLOM, Geoffrey (1981): "A Reexamination of the Role of Stocks in the Consumption Function and the Transmission Mechanism", *Journal of Money, Credit and Banking* 13, 215-220.

WORSLEY, K.J. (1979): "On the Likelihood-Ratio Test for a Shift in Location of Normal Populations", JASA 74, 365-367.

WU, D.M. (1973): "Alternative Tests of Independence between Stochastic Regressors and Disturbances", *Econometrica* 41, 733-750.

WU, D.M. (1974): "Alternative Tests of Independence Between Stochastic Regressors and Disturbances: Finite Sample Results", *Econometrica* 42, 529-546.

YAWITZ, Jess B., and Marshall, W.J. (1981): "Measuring the Effect of Callability on Bond Yields", *Journal of Money, Credit and Banking* 13, 60-71.

AUTHOR INDEX

SUBJECT INDEX